ibvt-Schriftenreihe

Schriftenreihe des Institutes für Bioverfahrenstechnik
der Technischen Universität Braunschweig

Herausgegeben von Prof. Dr. Christoph Wittmann

Band 62

**Cuvillier-Verlag
Göttingen, Deutschland**

Herausgeber
Prof. Dr. Christoph Wittmann
Institut für Bioverfahrenstechnik
TU Braunschweig
Gaußstraße 17, 38106 Braunschweig
www.ibvt.de

Bibliographische Informationen der Deutschen Nationalbibliothek
Die Deutsche Nationalbibliothek verzeichnet diese Publikation in der Deutschen Nationalbibliographie; detaillierte bibliographische Daten sind im Internet über *http://dnb.d-nb.de* abrufbar.
1. Aufl. – Göttingen: Cuvillier, 2011

© Cuvillier-Verlag · Göttingen 2011
    Nonnenstieg 8, 37075 Göttingen
    Telefon: 0551-54724-0
    Telefax: 0551-54724-21
    www.cuvillier.de

1. Auflage, 2011
Gedruckt auf säurefreiem Papier

ISBN 978-3-86955-970-4
ISSN 1431-7230

ibvt-Schriftenreihe Band 62

"Modelling the risk of chlorinated hydrocarbons in urban groundwater"

von

Tillman Greis

aus Lauterbach/Hessen

Schriftleiter/Herausgeber: Prof. Dr. Christoph Wittmann

© Cuvillier-Verlag – Göttingen 2011

Technische Universität Braunschweig

Von der Fakultät Architektur, Bauingenieurwesen und Umweltwissenschaften
der Technischen Universität Carolo-Wilhelmina zu Braunschweig
zur Erlangung des Grades eines Doktor-Ingenieurs (Dr.-Ing.)
genehmigte Dissertation

Eingereicht am	14. September 2011
Disputation am	03. November 2011
Berichterstatter	Prof. Dr. Andreas Haarstrick
	Prof. Dr. Enrica Caporali

Danksagung

Allen voran gilt ein besonderes Wort des Dankes Herrn Prof. Dr. Andreas Haarstrick, nicht nur für die Übernahme des Erstreferats, sondern auch für die vielen hilfreichen Diskussionen, die Ratschläge und das entgegengebrachte Vertrauen. Ohne diese Unterstützung wäre die Arbeit nicht möglich gewesen.

Frau Prof. Enrica Caporali möchte ich für die Übernahme des Zweitreferats danken. Ihre stets offene und zuvorkommende Art und Weise hat die Zusammenarbeit für mich um vieles leichter gemacht.

Auch Herrn Prof. Dr.-Ing. Joachim Stahlmann gilt mein Dank für die Übernahme des Prüfungsvorsitzes.

Bei Herrn Prof. Dr. Christoph Wittmann sowie Herrn Prof. a.D. Dr.-Ing. Dietmar Hempel möchte ich mich für die Möglichkeit bedanken, meine Arbeit am Institut für Bioverfahrenstechnik durchführen zu können.

Herrn Prof. Dr. Matthias Schöniger danke ich herzlich für zahlreiche fachliche Diskussionen, die gute Zusammenarbeit und so manche lehrreiche Lektion fürs Leben. Auch Frau Dipl.-Geoökol. Kathrin Helmholz gilt mein ganz besonderer Dank für die gute Zusammenarbeit im gemeinsamen Projekt.

Allen Mitgliedern des Internationalen Graduiertenkollegs 802 möchte ich meinen Dank für die schöne gemeinsame Zeit in Florenz (und natürlich auch in Braunschweig) ausdrücken. Dank euch durfte ich vieles aus den unterschiedlichsten wissenschaftlichen Disziplinen erlernen und die italienische Lebensweise in allen Facetten erfahren.

Darüber hinaus möchte ich den Mitarbeitern des Instituts für Bioverfahrenstechnik ganz herzlich danken für das allzeit gute Arbeitsklima, die gemeinsam verbrachten Tage (und Abende) und alles andere. Die schöne Zeit am Institut werde ich bestimmt nie vergessen. Danke dafür!

Zuletzt möchte ich meiner Familie für die Unterstützung danken, die mit Sicherheit am Wichtigsten für diese Arbeit war. Vielen Dank!

List of publications

Journal publications and book contributions:

- Greis, T.; Helmholz, K.; Schöniger, H.M. and Haarstrick, A.: "Modelling of spatial contaminant probabilities of occurrence of chlorinated hydrocarbons in an urban aquifer", Environmental Monitoring and Assessment, DOI:10.1007/s10661-011-2209-1, in press.

- Krujatz, F.; Haarstrick, A.; Nörtemann, B. and Greis, T.: „ Assessing the toxic effects of nickel, cadmium and EDTA on growth of the plant growth-promoting rhizobacterium *Pseudomonas brassicacearum* ", Water, Air & Soil Pollution, DOI:10.1007/s11270-011-0944-0 in press.

Conference contributions:

- Greis, T.; Helmholz, K.; Schöniger, H.M. and Haarstrick, A.: "Modelling reactive transport of chlorinated hydrocarbons in groundwater under spatially varying redox conditions", ModelCare 2011 conference, Leipzig (Germany), 19-22 september 2011; oral presentation

- Helmholz, K.; Schöniger, H.M.; Nienstedt, D. and Greis, T.: „Numerical stochastic simulation for risk assessment of an urban aquifer system", ModelCare 2011 conference, Leipzig (Germany), 19-22 september 2011; poster presentation.

Summary

The main target of this work is the development of a risk assessment approach of groundwater contaminations in an urban area. A multidisciplinary approach was performed to assess degradation potential of contaminant species in soil and the inherent risk posed by anthropogenic substances. Experimental and theoretical methods range from chemical and biochemical to geo-engineering applications. This especially concerns molecular biological, wet-chemistry techniques and Finite Element groundwater transport and reaction modelling. The spatial dimensions considered in the studies range from small laboratory to field scales. With regard to these applied methods the overall goal was a holistic risk approach to intertwine both, reactive transport of contaminants in groundwater and human health risks.

An experimental area polluted with chlorinated ethenes and located in Braunschweig was chosen to validate the model. Based on repeated measurement campaigns pollutant concentrations as well as other environmental parameters and chemical data were determined and their quality was valued at assessed literature data. With the obtained experimental data a groundwater reactive transport model was established and validated. Further, to test the influence of the flow, transport and reaction model parameters a sensitivity analysis was performed. Based on this analysis, contaminant specific probabilities of occurrence were calculated with a Monte-Carlo simulation approach. Additionally, model optimisation techniques were applied to improve conformity of simulated and experimental data.

It could be shown, that a simplified first-order reaction kinetic is only partially capable of rendering the measured field data. Hence, it was necessary to extend and improve the underlying degradation kinetics by means of modified Monod-equations. Here, the extension mainly comprises the introduction of inorganic electron acceptors as well as inhibiting reactions to refine the reductive dechlorination process of the chlorinated hydrocarbons. The next important step concerns the derivation of a health risk approach from the aforementioned concentration probabilities of occurrence. The combined model approach finally enables to calculate spatial and temporal health risk occurrence.

Zusammenfassung

Ziel dieser Arbeit ist die Risikoermittlung einer Altlast in einem urbanen Umfeld. Ein multidisziplinärer Ansatz zur Beurteilung des Abbaupotentials von Schadstoffen im Grundwasser wurde gewählt, um das Gesamt-Risikopotential abschätzen zu können. An experimentellen und theoretischen Methoden wurde dabei auf chemische und biochemische Techniken bis hin zu geowissenschaftlichen Untersuchungen zurückgegriffen. Im Speziellen kamen molekular-biologische und nasschemische Analysen sowie die Finite-Elemente Modellierung von Transport und Reaktion im Grundwasser zur Anwendung. Die räumlichen Dimensionen reichten dabei von kleinskaligen Laborexperimenten bis hin zu Messungen auf der Feldskala. Mit Hilfe dieses ganzheitlichen Ansatzes war es möglich, die von Altlasten ausgehenden Gefahren mit potentiellen Gesundheitsschäden von Anwohnern zu verknüpfen.

Als Beispiel wurde ein mit chlorierten Ethenen belastetes Gebiet im Raum Braunschweig ausgewählt, auf welches das hier entwickelte Modell angewendet wird. Während wiederholter Messkampagnen wurden Schadstoffkonzentrationen sowie verschiedene Umweltparameter ermittelt und mit vorhandenen Daten in der Literatur verglichen. Anhand der Messdaten wurde ein Grundwassermodell zur Simulation von Fließ-, Transport- und Reaktionsprozessen erstellt und validiert. Mittels einer Sensitivitätsanalyse konnten daraufhin wichtige Transport- und Reaktionsparameter ermittelt werden, die anschließend mithilfe einer Monte-Carlo Simulation zur Berechnung von schadstoffspezifischen räumlichen Auftretenswahrscheinlichkeiten genutzt wurden.

Es zeigte sich während der Anwendung der beschriebenen Methoden, dass ein auf der vereinfachten linearen Reaktionskinetik beruhender Schadstoffabbau die realen Verhältnisse nur teilweise beschreiben konnte. Daher wurde ausgehend von den vorliegenden Daten ein Reaktionsmodell auf Basis der Monod-Kinetik entwickelt. Ferner wurden anorganische Elektronenakzeptoren als inhibierende Substanzen für die reduktive Dechlorierung von Chlorethenen implementiert und dieses erweiterte kinetische Modell auf die Fallstudie angewendet. In einem weiteren Schritt wurden die bereits erwähnten Schadstoff-Auftretenswahrscheinlichkeiten mit einem Gesundheitsrisikoansatz verknüpft. Abschließend konnte damit eine Risikoanalyse durchgeführt werden, um räumliche und zeitliche Gefährdungspotentiale abzuschätzen.

Contents

1 Introduction and scope of work 1

2 Theoretical background 4

 2.1 Chlorinated ethenes as groundwater contaminants..............................4

  2.1.1 Physical and chemical properties of chlorinated ethenes 4

  2.1.2 Chlorinated ethene degradation in the environment......................5

 2.2 Groundwater flow, transport and reaction modelling...........................9

  2.2.1 Advection...9

  2.2.2 Dispersion..9

  2.2.3 Diffusion...10

  2.2.4 Adsorption 11

  2.2.5 Degradation 13

  2.2.6 Combined differential transport equations...........................16

  2.2.7 Boundary conditions 17

 2.3 Risk assessment approach for groundwater contaminations 18

  2.3.1 Classification of brownfield sites 18

  2.3.2 Risk management concepts..20

  2.3.3 Health risk assessment approaches 22

3 Experimental area..26

 3.1 Geological setting and contaminant release 26

 3.2 Field measurements...28

4 Laboratory and soil column experiments...34

 4.1 Adsorption and retardation soil column experiment 34

  4.1.1 Materials and methods...34

  4.1.2 Results of the experiment 35

 4.2 Batch experiment of chlorinated ethene degradation..........................37

  4.2.1 Materials and methods...37

  4.2.2 Results of the experiment 39

  4.2.3 Modelling of the batch test experiment..............................42

 4.3 Characterisation of chloroethene-degrading bacteria 45

  4.3.1 Materials and methods...45

  4.3.2 Results of the experiment 47

5   Finite Element groundwater modelling 50

   5.1   Steady-state Finite Element model...51

      5.1.1   Definition of model domain and initial conditions...............................51

      5.1.2   Setup and calibration of the hydraulic model.....................................53

      5.1.3   Setup and calibration of the transport and reaction model..................54

   5.2   Model results and model optimisation 56

      5.2.1   Sensitivity analysis...56

      5.2.2   Monte-Carlo simulation for calculation of probability isolines.............58

      5.2.3   Model optimisation using Monte-Carlo method 61

   5.3   Development of a redox-dependant degradation kinetic..................................67

      5.3.1   Formulation of the problem and conceptual approach........................67

      5.3.2   Monod-based model using electron-acceptor inhibition kinetic..........70

      5.3.3   Setup of a transect model 75

6   Risk assessment approach 80

   6.1   Background and model assumptions..80

   6.2   Modelling health risk...83

      6.2.1   Cancer risk approach...83

      6.2.2   Hazard Quotient approach 88

7   Summary and discussion...93

8   Outlook and future prospects 97

9   Nomenclature 99

10   References 103

11   Appendix..110

List of figures

Figure 2.1: Chlorinated ethene degradation pathway 5
Figure 2.2: Oxidation numbers exemplarily shown for the reaction of PCE to TCE; reduction can be observed at the right-hand carbon-atom 6
Figure 2.3: Redox reactions catalysed by microorganisms and the associated redox potential at pH 7, redrawn after Schwoerbel (1987) 8
Figure 2.4: Processes leading to hydrodynamic dispersion 10
Figure 2.5: Distribution of degradation rate constants (first-order) for chlorinated ethenes, redrawn after (Wiedemeier et al., 1998) for PCE, TCE, VC; own literature assessment for cis-DCE 15
Figure 2.6: Proposed probabilistic risk management framework for soil contaminations, redrawn and extended after Pliefke et al. (2007) 19
Figure 3.1: Investigation area with model domain and observation wells; horizontal line represents a petrographic cross-section shown in Figure 5.2 26
Figure 3.2: Interpolated contaminant plume at the case study area "Schützenplatz" for PCE 29
Figure 3.3: Interpolated contaminant plume at the case study area "Schützenplatz" for TCE 29
Figure 3.4: Interpolated contaminant plume at the case study area "Schützenplatz" for cis-DCE 30
Figure 3.5: Interpolated contaminant plume at the case study area "Schützenplatz" for VC 30
Figure 3.6: Interpolated nitrate concentrations at the Schützenplatz area based on measurement data 32
Figure 3.7: Interpolated dissolved iron concentrations at the Schützenplatz area based on measurement data 33
Figure 4.1: Experimental setup and photo of the laboratory soil column 34
Figure 4.2: Measured (dotted) and simulated (solid) normalised concentrations /conductivity of the soil column experiments to determine transport behaviour of a conservative tracer (NaCl, shown as conductivity), PCE and cis-DCE 35
Figure 4.3: Soil column experiment showing heavy metal desorption with EDTA; comparison of experimental and simulation results 37
Figure 4.4: Chlorinated ethene concentrations in the different batch tests: a) reference sample; b) EHC sample; c) glucose sample; d) nitrate sample 39
Figure 4.5: Nitrate and sulphate concentrations for the batch experiments: a) reference sample; b) EHC sample; c) glucose sample; d) nitrate sample 40
Figure 4.6: Approximated conversion rates of chlorinated ethene degradation in the batch experiment 41
Figure 4.7: Simulation results (sim) and measurement data (meas) of the batch test: a) reference sample; b) EHC sample; c) glucose sample; d) nitrate sample 44
Figure 4.8: Observation wells along the transect line in the Schützenplatz area sampled for detection of dechlorinating microorganisms 46
Figure 4.9: Gel-photo of the PCR with 774f/1172rmod-primers 48
Figure 4.10: Measurement data for chloroethene concentrations and redox potential at the observation wells sampled for the characterisation of dehalogenating bacteria 48
Figure 5.1: Conceptual approach for calculation of contaminant-specific probabilities of occurrence for risk assessment purposes 50
Figure 5.2: Exemplary illustration of the layer model implemented by means of petrographic data (vertically exaggerated); cross-section shown here according to horizontal line in Figure 3.1 51

Figure 5.3: Model domain with FE mesh and boundary conditions................................52

Figure 5.4: Water balance of the steady-state groundwater model (positive values: inflow; negative values: outflow) 53

Figure 5.5: Scatter plot of measured versus simulated groundwater head (referring to meters above mean sea level (m AMSL))..54

Figure 5.6: Scatter plot of measured versus simulated concentrations................................55

Figure 5.7: Comparison between variation of diffusion coefficient D_d (left) and hydraulic conductivity K_F (right) on transport of PCE (Greis et al., 2011)................................57

Figure 5.8: Correlation analysis to derive a sufficient number of MC simulations (Greis et al., 2011) 59

Figure 5.9: Isoline visualisation of the compound specific probabilities of occurrence, referring to 50 years simulation time (Greis et al., 2011)................................60

Figure 5.10: Isoline visualisation of the compound specific probabilities of occurrence, referring to 70 years simulation time (Greis et al., 2011)................................61

Figure 5.11: Development of cumulated normalised error Err_{norm} (a) and cumulated absolute error Err_{abs} (b) in the MC simulations 63

Figure 5.12: Comparison of absolute error development of the four simulated species (a: PCE, b: TCE, c: cis-DCE, d: VC) 64

Figure 5.13: Parameter development in the MC simulations normalised to reference model (1: Reference model, 2-5: MC simulations 1-4)................................65

Figure 5.14: Comparison of concentration isolines derived from instationary (dashed lines) and steady-state (solid lines) groundwater model of the Schützenplatz area 67

Figure 5.15: Top view of the transect (black line) with observation wells and particle tracking (blue lines)................................69

Figure 5.16: concentration distribution of chloroethenes (a) and selected electron acceptors (b) on the transect shown in Figure 5.15................................69

Figure 5.17: Comparison of the cumulated total errors for the model optimisation via MC approach (chapter 5.2.3) and the cumulated total error for the inhibition model 71

Figure 5.18: Comparison of measured and simulated contaminant concentrations of the redox model referring to chlorinated hydrocarbons 72

Figure 5.19: Illustration of model results using the simplified redox kinetic applied to the case study area for PCE 73

Figure 5.20: Illustration of model results using the simplified redox kinetic applied to the case study area for TCE 73

Figure 5.21: Illustration of model results using the simplified redox kinetic applied to the case study area for cis-DCE................................74

Figure 5.22: Illustration of model results using the simplified redox kinetic applied to the case study area for VC................................74

Figure 5.23: Overview of the developed microbially mediated reductive dechlorination model................................76

Figure 5.24: concentration development of the transect model using the holistic reductive dechlorination approach 78

Figure 5.25: Comparison of the results for the model scale-up towards the case study model: parameter set of the 2D transect model implemented into the 3D case study model 79

Figure 6.1: Overview of   the model domain with location of contaminant source area (in red, south-east), assumed residents' groundwater wells (blue dots, north-west) and calculated cis-DCE distribution to illustrate the intrinsic risk situation................................82

Figure 6.2: Cancer risk rate development with upper and lower bounds for transport model uncertainty and population variability exemplarily shown for one groundwater well in the case study area (see Figure 6.1) 84

Figure 6.3: cumulated cancer risk evolution with upper and lower bounds for model and population uncertainty (10% and 90% quantile, respectively) exemplarily shown for one observation well in the study area (see Figure 6.1).. 85

Figure 6.4: Cancer risk for current time period (50 years simulation time); a: probability distribution for exceedance of a risk level of 0.001 cumulated over all chloroethene compounds for the population median; b: uncertainty evaluation using fixed aquifer model uncertainty and varying population quantiles at a risk level of 0.001 87

Figure 6.5: Temporal evolution of median annual Hazard Quotient according to equation (2.19) with upper and lower bounds for model and population uncertainty (10% and 90% quantile) for an example well in the study area 88

Figure 6.6: Exceedance of Hazard Quotient (HQ > 1 according to equation (2.19)) at 50 years simulation time with regard to model uncertainty; a) PCE, b) TCE, c) cis-DCE, d) VC 90

Figure 6.7: Exceedance of Hazard Quotient for each chloroethene compound (at 50 years simulation time) with regard to population uncertainties, a) PCE, b) TCE, c) cis-DCE, d) VC 91

Figure 11.1: Adsorption experiment and simulation of chloroethenes in the soil column showing all 5 species (see chapter 4.1) 111

Figure 11.2: Dirichlet boundary condition used for infiltration of PCE 124

Figure 11.3: Scatter plot of measured and simulated water table of the preliminary reactive transport model... 125

Figure 11.4: Concentration of electron acceptors over the transect length implemented in the transect model; fixed values (not shown) for oxygen (0.1 $mg \cdot l^{-1}$) and sulphate (400 $mg \cdot l^{-1}$) 128

List of tables

Table 2.1: Physicochemical properties of chlorinated ethenes (according to Schwarzenbach et al., 2003 and ChemIDplus Database) 4

Table 2.2: Redox reactions with corresponding Gibbs free energy $\Delta G°$ and standard reduction potential $E°$ (according to Dolfing et al. , 2006)... 8

Table 2.3: Empirical equations for calculation of soil partition coefficients K_D according to different authors.. 12

Table 2.4: Octanol-water partition coefficients for different chloroethene compounds 13

Table 2.5: Short overview over published first-order degradation rates for chlorinated ethenes $[y^{-1}]$ 14

Table 2.6: Monod parameters for chlorinated ethene degradation derived from literature 16

Table 2.7: Summary of cancer slope factors (CSF) and reference doses (RfD) for chlorinated ethenes 25

Table 3.1: Soil layers present in the case study area (soil type by analysis of drilling profiles, hydraulic conductivity derived from literature, organic carbon fraction determined by own measurements) 27

Table 3.2: mean measurement values and standard deviation for exemplarily chosen observation wells 31

Table 4.1: Sorption parameters determined by experiment and simulation 36

Table 4.2: Setup of the batch experiment for assessment of chloroethene degradation................. 38

Table 4.3: Calculated first-order degradation rates for TCE in selected experiments 42

Table 4.4: Summary of reaction rates and inhibition constants used in the batch test simulations 43

Table 5.1: soil parameters used in the base model and standard deviation σ of parameters applied for the MC simulation 58

Table 5.2: contaminant parameters used in the base model and standard deviation σ of the parameters generated for the MC simulation 58

Table 6.1: Contribution of single chemical compounds to the cancer risk at the example well 86

Table 6.2: Contribution of single chemical compounds to the Hazard Quotient at the example well 89

Table 11.1: General column layout parameters for adsorption models 110

Table 11.2: Temporal and control data for adsorption models 110

Table 11.3: General flow data for adsorption models 110

Table 11.4: Transport and sorption data for adsorption models 111

Table 11.5: Primer design for the PCR with groundwater samples 113

Table 11.6: Composition of the PCR master mix 114

Table 11.7: Chlorinated ethene degradation and stoichiometric matrix 115

Table 11.8: Competing redox reactions involved in chlorinated ethene degradation 116

Table 11.9: Brownfield site statistics for Germany according to Umweltbundesamt (Date: 2010) (as discussed in chapter 2.3.1) 117

Table 11.10a: Measurement data from groundwater wells including measurement uncertainty .. 118

Table 11.11: General settings of the FE model "Schützenplatz" 123

Table 11.12: Temporal and control data of the FE model "Schützenplatz" 123

Table 11.13: Hydraulic parameters of the reference model (chapter 5.1) and range of stochastic variation for the MC simulation (chapter 5.2) 124

Table 11.14: Species related parameters of the reference model (chapter 5.1) and stochastic variation for the MC simulation (chapter 5.2) 124

Table 11.15: Hydraulic parameters used in the redox-dependent degradation model (chapter 5.3.2) 125

Table 11.16: Species related parameters used in the redox-dependent degradation model (chapter 5.3.2) 126

Table 11.17: Inhibition constants used in the redox-dependent degradation model (chapter 5.3.2) 126

Table 11.18: General model layout for transect model simulations 127

Table 11.19: Temporal and control data for transect model simulations 127

Table 11.20: General flow data for adsorption models 127

Table 11.21: Transport and sorption data for adsorption models 128

Table 11.22: Inhibition constants of PCE degradation by different electron acceptors 129

Table 11.23: Inhibition constants of TCE degradation by different electron acceptors 129

Table 11.24: Inhibition constants of cis-DCE degradation by different electron acceptors 129

Table 11.25: Inhibition constants of VC degradation by different electron acceptors 129

Table 11.26: Growth and death rate for different microorganisms; growth inhibition constants for different electron acceptors 130

1 Introduction and scope of work

Anthropogenic substances like chlorinated ethenes have been released into the environment in countless sites worldwide, especially in industrialised countries. The release of such contaminants was mainly caused due to the lack of environmental restrictions during 1950-1970 and primarily affects soil and groundwater ecosystems. The European Environmental Agency (EEA) estimates the number of sites requiring cleanup to exceed 100,000 in European countries (European Environment Agency, 2005). In Germany around 320,000 sites have been identified as suspicious areas for soil or groundwater contaminations according to the Bundes-Bodenschutzgesetz (BBodSchG) and are thus catalogued in a register for contaminated sites. The categorisation as brownfield site typically requires an unambiguous prove of contamination occurrence accompanied with an impact on public safety originating from the area of concern. Over 13,500 sites nationwide have already been classified as brownfield sites, meaning that pollutant concentrations have verifiably exceeded regulatory values and hence remediation measures are legally stipulated. However, only around 4,000 are currently subjected to clean-up activities (Umweltbundesamt, 2010).

Typically the low degradation rates of many recalcitrant substances result in a long-term impact on affected groundwater and soil ecosystems: pollutants often persist in soils for decades and pose danger to environment and residents, respectively. Necessary remediation actions conducted at polluted sites often require plenty of time and normally do not completely remove all contaminant substances from the affected site. In particular for wide-spread contamination plumes, financial resources for remediation campaigns often suffice merely for clean-up of the core area, leaving major parts of polluted areas untreated.

During so called Monitored Natural Attenuation (MNA) approaches the future development of affected sites is frequently tried to be modelled in order to assess the fate of contaminants during groundwater transport. The main objective of this approach is to estimate the self-healing potential of the ecosystem of concern and to calculate the potential risks arising from chemical substances in soil and groundwater for different pathways and receptors. However, uncertainties in parameters and inhomogeneities of the system often lead to highly variable model results, which are difficult to interpret for decision makers. In particular, problems with modelling often arise for sites polluted with chlorinated ethenes due to their sequential degradation chain. Their desirable properties (volatile, highly stable, non-flammable, cheap) have led to extensive usage as solvents and degreasing agents in dry-cleaning processes, for instance in laundries, metalworking or automotive industries. Hence, chloroethenes like tetrachloroethene (PCE) are common organic contaminants at many polluted sites. Enquiries estimate that a percentage of over 60% of all brownfield sites are contaminated with this kind of pollutants (Stupp et al., 2007). Generally chlorinated ethenes are subject to degradation processes in aquifer ecosystems leading to a spectrum of conversion products found at contaminated areas. These secondary products besides PCE typically are trichloroethene (TCE), cis-1,2-dichloroethene (cis-DCE) and vinyl chloride (VC), which all possess toxic and partially even carcinogenic properties and therefore are of concern for human health if taken up. Health-impairing and carcinogenic effects of these substances have been demonstrated lately and have led to environmental restrictions in usage and disposal.

In consequence, industrial sites contaminated with Dense Non-Aqueous Phase Liquids (DNAPLs) pose a high potential danger for the environment and humans especially in inhabited urban areas. Hence, the improvement of existing risk assessment tools for contaminated sites on the field-scale appears to be of avail in order to predict future development of contaminated aquifers and to lower the risk for affected residents as well as boundary ecosystems. In order to provide an appropriate risk analysis, the construction of a calibrated groundwater flow, transport and reaction model is necessary for demonstration of feasibility of the developed tools.

An experimental area polluted with chlorinated ethenes located in Braunschweig was chosen to evaluate and review the outcomes of the developed groundwater model and to calculate the potential risk for residents exposed to groundwater contaminations in an urban area. This work is targeted on the assessment and the grading of risks associated with groundwater contaminations and to demonstrate the feasibility of the approach for the case study mentioned. Additionally, the area of concern is currently under consideration for a spatially confined in-situ remediation measure.

Further, this work aims to develop modelling tools enabling an improved prediction of contaminant fate in groundwater under spatially varying environmental conditions, e.g. redox potential, which is identified to lead to better understanding and planning of the respective clean-up activities. For all chlorinated ethenes reaction rate constants are generally low and hence, degradation occurs very slowly. In contrast to other problematic organic substances (e.g. PAH and BTEX), adsorption onto soil particles is rather low, resulting in widespread contamination plumes in comparison to the other pollutants (Stupp and Paus, 1999), which turn chlorinated ethene remediation measures into a challenging task. In addition, environmental parameters like redox potential have been proven to show a significant influence on contaminant degradation. However, literature dealing with this topic is not very comprehensive (Widdowson, 2004; Doong et al., 1996). The development of a sophisticated method to predict pollutant decay under spatially varying environmental conditions seems to be advisable and is also substantial in this work.

The work presented here mainly focuses on uncertainties of transport parameters and the establishment of an improved chemical model of chlorinated ethene degradation for a more accurate calculation of environmental risk. The conclusions of these evaluations are used in a risk assessment approach applied to a real-world field-scale scenario in order to calculate health risks arising from groundwater contaminations. Conducted in collaboration of the Technische Universität Braunschweig (Germany) with the University of Florence (Italy), this thesis is embedded in the framework of the International Graduate College 802 "Risk Management of Natural and Civilization Hazards on Buildings and Infrastructures", which focuses in a multidisciplinary manner on risk assessment and risk management of different hazards.

2 Theoretical background

2.1 Chlorinated ethenes as groundwater contaminants

2.1.1 Physical and chemical properties of chlorinated ethenes

Chlorinated ethenes are a group of chemical compounds based on the ethene molecule, which is substituted with a varying number of chlorine atoms. This group consists of the species tetrachloroethene (PCE), trichloroethene (TCE), cis-, trans-, and 1,1-dichloroethene (DCE) and finally vinyl chloride (VC). They all share similar physicochemical properties such as high density, low aqueous solubility and a low degradability in the environment. The degree of substitution defines some of their specific properties: an increasing number of chlorine substituents leads to an increase in density, melting point and boiling point (see Table 2.1).

Table 2.1: Physicochemical properties of chlorinated ethenes (according to Schwarzenbach et al., 2003 and ChemIDplus Database)

Compound	Molar mass [g·mol^{-1}]	Density [g·cm^{-3}]	Melting point [°C]	Boiling point [°C]	Log K_{OW} [-]	Solubility in water [mg·l^{-1}]
PCE	165.83	1.62	-22.4	121.1	3.4	206
TCE	131.39	1.46	-73.0	87.0	2.42	1280
cis-DCE	96.95	1.27	-81.0	60.0	1.86	6410
VC	62.5	0.91*	-153.8	-13.7	1.62	8800

*at boiling point

All chlorinated ethenes are hydrophobic substances, but with decreasing number of chlorine substituents the solubility in water increases. This is also reflected by the octanol-water partition coefficient (Log K_{OW}), which summarises the potential of a substance to dissolve in water or organic solvents (in particular n-octanol). This property leads to a higher mobility of lower-chlorinated ethenes in soil in comparison to higher-substituted ones. Besides their position in the degradation chain (see chapter 2.1.2) this is one reason why cis-DCE and VC are found in the downstream part and PCE and TCE mainly in the upstream part of affected groundwater systems. The high density of all chlorinated ethenes results in settling of the solvent phase downwards onto impervious soil layers (aquitard) during the source release period. Originating from the solvent phase a delayed dissolution of chloroethene compounds into the groundwater (caused by the low aqueous solubility) often leads to a long-lasting impact onto the aquifer.

Toxicologically all chlorinated ethenes pose harmful effects towards humans. These effects manifest themselves in toxic or carcinogenic properties. Thus, guideline values for environmental occurrence were established which in particular are 10 $\mu g \cdot l^{-1}$ in drinking water according to Trinkwasserverordnung 2001 (cumulated concentration for all chloroethenes), 100 $\mu g \cdot m^3$ for indoor air according to Bundes-Immissionsschutzverordnung 1990 (guideline value for PCE) and 10 $\mu g \cdot l^{-1}$ for soil concentrations according to Bundes-Bodenschutzverordnung 1998 (inspection value for PCE).

2.1.2 Chlorinated ethene degradation in the environment

The main reaction pathway for degradation of chlorinated ethenes is described by a simple consecutive reaction chain. The reaction chain typically starts with the compound tetrachloroethene (PCE) and proceeds with trichloroethene (TCE), cis-1,2-dichlorothene (cis-DCE), vinyl chloride (VC) towards ethene (eth) (see Figure 2.1). In each step of the degradation reaction hydrogen is consumed and hydrochloric acid is released. Although there are other possible isomers for formation of dichloroethene (1,1-DCE; trans-1,2-DCE), mainly cis-1,2-dichloroethene (cis-DCE) is built by microbiologically mediated degradation processes (Bradley, 2000). Each step of the degradation chain is dependant on several environmental conditions like pH-value, temperature and redox conditions and can be described by an own reaction equation (Clement et al., 1998).

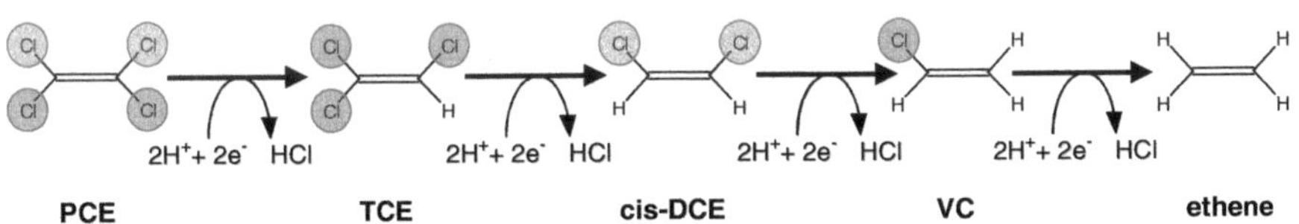

Figure 2.1: Chlorinated ethene degradation pathway

Degradation of chlorinated hydrocarbons typically depends on the environmental conditions predominating in the polluted area and occurs naturally in affected aquifer systems (Mulligan and Yong, 2004). Besides abiotic reactions, two important metabolic pathways are identified for mineralisation of chloroethene pollutants. The first is called "aerobic cometabolisation", during which dehalogenation occurs as a side-reaction of growth substrate catabolism (Alvarez-Cohen and Speitel, 2001), e.g. toluene (Devlin et al., 2004) or methanol (Yang et al., 2008). Dissolved oxygen is typically used as electron acceptor in this case.

The other important degradation pathway is called "reductive dechlorination" (Bradley and Chapelle, 2010; Olaniran et al., 2004), where electrons are transferred from an organic substrate (e.g. glucose) towards an electron-accepting substance. This pathway is normally used by particular microorganisms to get rid of excess electrons originating from other catabolic reaction pathways (e.g. glycolysis) (Schwarzenbach et al., 2003). On the one hand, inorganic substances like oxygen, nitrate or sulphate might serve as electron acceptors; on the other hand chlorinated ethenes are capable of consuming electrons and are being subjected to dechlorination in this step. In total the average oxidation number of carbon atoms in the chloroethene molecule is reduced in each reaction step (Figure 2.2).

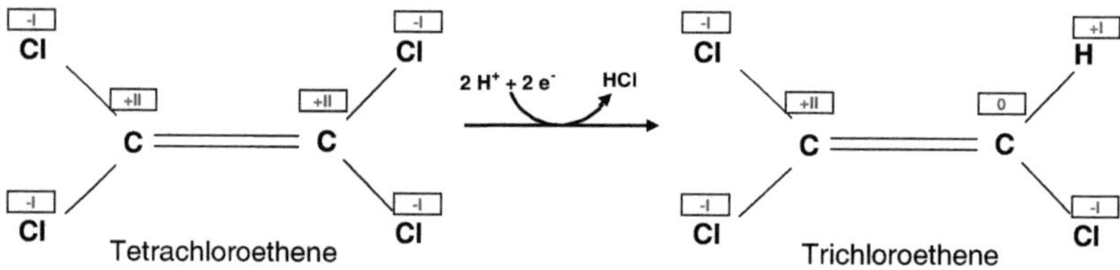

Figure 2.2: Oxidation numbers exemplarily shown for the reaction of PCE to TCE; reduction can be observed at the right-hand carbon-atom

The reductive dechlorination reaction is mediated by a specific group of soil microorganisms, which utilise a set of enzymes called reductive dehalogenases for this task (Maillard et al., 2011). Typically, most of these enzymes are only capable of catalysing the reaction from PCE to cis-DCE. To date, only one species is known to fully degrade PCE to ethene: the strain *Dehalococcoides ethenogenes* 195 (Yan et al., 2009). Polluted sites, where these bacteria are found, normally tend to show a complete mineralisation of chloroethenes, while sites without this specific strain often exhibit an accumulation of cis-DCE, but no further breakdown products (Hendrickson et al., 2002).

From the thermodynamic point of view, reductive dehalogenation in soil and groundwater ecosystems competes with reduction of a variety of inorganic electron acceptors. These electron acceptors are presumed to inhibit dehalogenation of chloroethenes according to their position in the redox chain (see Figure 2.3). The dechlorination reaction does only proceed under thermodynamically favourable conditions. Reactions with higher energy yield, i.e. a more negative Gibbs free energy, are preferred in comparison to those with a lower energy yield or more positive Gibbs free energy, respectively (compare Table 2.2). The redox potential of the groundwater system thus is a crucial indicator for the assessment of the pollutant degradation potential (Christensen et al., 2000).

For each redox couple a standard reduction potential can be calculated by the Nernst equation (equation (2.1)).

$$E = E^0 + \frac{R \cdot T}{z_e \cdot F} \ln \frac{a_{ox}}{a_{red}}$$

(2.1)

E: reduction potential

E^0: standard reduction potential

R: universal gas constant (R = 8.314472 $J \cdot K^{-1} \cdot mol^{-1}$)

T: absolute temperature

z_e: number of transferred electrons

F: Faraday constant (F = 9.648533$\cdot 10^4$ $C \cdot mol^{-1}$)

a_{ox}, a_{red}: activity of relevant species of oxidant and reductant

In aquifer systems temperature differences are assumed to be negligible and thus, the resulting reduction potential of a redox reaction is mainly dependant on the respective standard reduction potential E°and the concentration of oxidised and reduced species.

Additionally, the Gibbs free energy for each reaction is associated with the standard reduction potential by means of the following equation (2.2):

$$\Delta G^0 = -z_e \cdot F \cdot E^0$$

(2.2)

ΔG^0: Gibbs free energy

As a consequence, the redox potential of an aquifer system is supposed to be a mixed potential regarding the participating redox couples, i.e. in unaffected groundwater systems the redox potential is a result of the concentration of inorganic electron acceptors present in the system. Given the fact that sufficient carbon source is available, the electron acceptor with the highest energy yield is favoured to be transformed by microorganisms and hence, reduced by redox reactions during microbial metabolism. This leads to a sequential electron acceptor depletion starting from oxygen and ending with sulphate reduction and methane fermentation (Scheffer et al., 2008). This process is accompanied by a shift in redox potential (Figure 2.3).

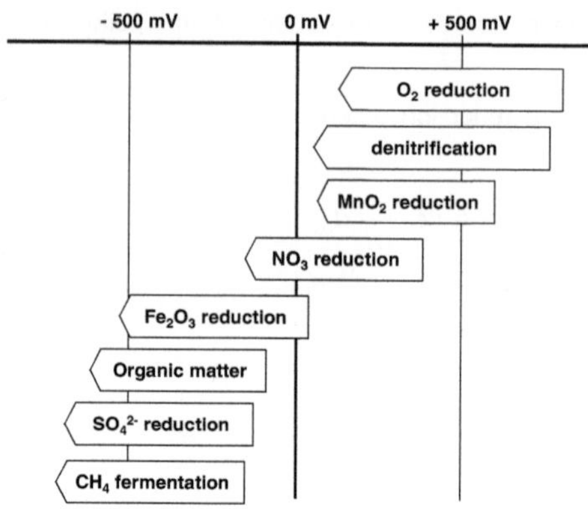

Figure 2.3: Redox reactions catalysed by microorganisms and the associated redox potential at pH 7, redrawn after Schwoerbel (1987)

The corresponding standard reduction potentials for chloroethene degradation are available in literature (Dolfing et al., 2006; Doong et al., 1996). Table 2.2 summarises the Gibbs free energy and standard reduction potential for the reduction of inorganic electron acceptors and chlorinated ethenes. It becomes obvious, that the reductive dehalogenation of chloroethenes is located in the same redox range as the reduction of inorganic electron acceptors and thus, from a thermodynamical point of view a competitive behaviour can be assumed (Widdowson, 2004). The theoretical energy yield from the reduction of chlorinated ethenes increases with their number of substitutes. Referring to Table 2.2 it becomes obvious that different degradation reactions are thermodynamically favoured under certain environmental conditions, e.g. the degradation of cis-DCE is energetically preferred after the depletion of nitrate, while PCE is degraded theoretically also in presence of this substance.

Table 2.2: Redox reactions with corresponding Gibbs free energy $\Delta G°$ and standard reduction potential E°(according to Dolfing et al. , 2006)

Reaction	ΔG^0 [kJ/electron]	E^0 [mV]
$O_2 + 4H^+ + 4e^- \rightarrow 2\,H_2O$	-78.7	816
$Fe^{3+} + e^- \rightarrow Fe^{2+}$	-74.4	771
$MnO_2 + HCO_3^- + 3H^+ + 2e^- \rightarrow MnCO_3 + 4H_2O$	-58.9	610
$NO_3^- + 2H^+ + 2e^- \rightarrow NO_2^- + H_2O$	-41.7	432
$SO_4^{2-} + 9H^+ + 8e^- \rightarrow HS^- + 4H_2O$	+20.9	-217
$PCE + H^+ + 2e^- \rightarrow TCE + Cl^-$	-55.4	574
$TCE + H^+ + 2e^- \rightarrow DCE + Cl^-$	-53.1 to -50.9*	550 to 527*
$DCE + H^+ + 2e^- \rightarrow VC + Cl^-$	-40.6 to -38.3*	420 to 397*
$VC + H^+ + 2e^- \rightarrow ethene + Cl^-$	-43.4	450

*depending on species of DCE involved (cis-, trans-, 1,1-DCE)

2.2 Groundwater flow, transport and reaction modelling

2.2.1 Advection

The flow of a fluid through a porous medium like soil is characterised by a constitutive equation called Darcy's equation. It describes the proportional relationship between the instantaneous discharge rate through the medium, the pressure drop over a certain distance and the viscosity of the fluid.

For modelling purposes the fluid motion is defined by the following equation (Diersch, 2009):

$$q = -K_r(s) \cdot K_f (\nabla h + \chi e) \tag{2.3}$$

q: Darcy flux vector;

$K_r(s)$: relative hydraulic conductivity ($0 < K_r \leq 1$, $K_r = 1$ if saturated at $s = 1$);

K_f: tensor of hydraulic conductivity for the saturated medium (anisotropy);

χ: buoyancy coefficient including fluid density effects;

e: gravitational unit vector;

h: hydraulic (piezometric) head

The variable K_f in equation (2.3) is generally known as hydraulic conductivity. It mainly depends on the physics and material properties of the porous medium, e.g. the grain size distribution of soil. In general, smaller grain sizes come along with a lower hydraulic conductivity. In case of saturated conditions and negligence of fluid density effects, fluid flux is only dependant on hydraulic conductivity and the existing gradient. The parameter range extends from 10^{-2} m·s^{-1} (highly permeable soil, e.g. gravel) and 10^{-9} m·s^{-1} (nearly impermeable soil, e.g. clay), typical values for soil lie in between 10^{-3} m·s^{-1} (sand) and 10^{-6} m·s^{-1} (silt) (Scheffer et al., 2008).

2.2.2 Dispersion

The hydrodynamic dispersion is an empirical factor which describes particle scatter in the fluid flow through a porous medium. On the microscopic scale the mechanism is determined by the distribution of flow velocity in soil pores and the pathway of each particle in the fluid around soil particles (see Figure 2.4). This results in a stochastically distributed length and width of flow paths which can be expressed by a dispersion coefficient for each spatial direction.

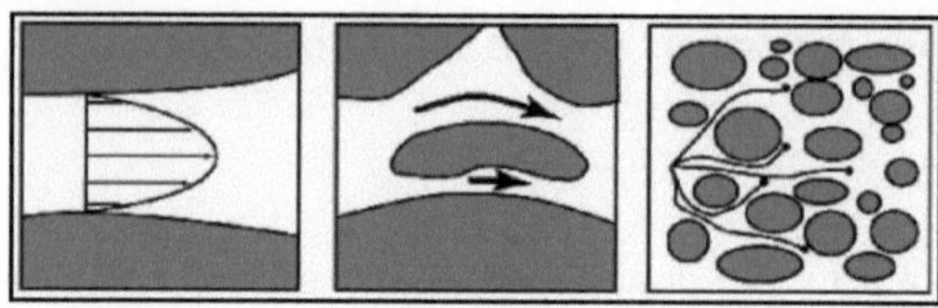

Figure 2.4: Processes leading to hydrodynamic dispersion

On the macroscopic scale, dispersion can be described by an (apparent) dispersion coefficient, which is often assumed to be scale dependent due to heterogeneities of the aquifer material. Dispersion can be distinguished between longitudinal, transversal and vertical dispersion. However, vertical dispersion is often neglected in groundwater modelling, e.g. in the Bear-Scheidegger dispersion equation only longitudinal and transversal dispersion are taken into account and the vertical dispersion is approximated with the transversal dispersion coefficient (Diersch, 2009). For modelling purposes the following formulation is made use of:

$$D_k = \left(\varepsilon D_{d_k} + \alpha_T \|q\|\right)I + (\alpha_L - \alpha_T)\frac{q \otimes q}{\|q\|}$$

(2.4)

D_k: tensor of hydrodynamic dispersion of species k;

D_{dk}: coefficient of molecular diffusion of species k;

I: unit tensor;

k: species indicator, k = 1,..., n;

ε: volume fraction

q: Darcy flux

α_L, α_T: longitudinal and transverse dispersivity of porous medium

As a rule of thumb, longitudinal dispersivity in aquifer systems is assumed to be one order of magnitude larger than transversal dispersivity. However, experimental studies on dispersion are rare. Often this parameter is estimated by the scale of the model and other soil parameters and ranges over four orders of magnitude (from below 0.1m to over 1000m) (Gelhar et al., 1992).

2.2.3   Diffusion

Molecular diffusion is generally defined as mass flux along concentration gradients due to Brownian motion. It is proportional to the squared velocity of the diffusing particles, which is dependent on temperature, viscosity of the fluid, and particle size. Three-dimensional diffusion is described by Fick's second law, which intertwines a relation between temporal and spatial concentration differences and thus, enables the calculation of instationary diffusion:

$$\frac{\partial C_k}{\partial t} = -D_{d_k} \nabla^2 C_k \qquad (2.5)$$

C_k: concentrations of species k

D_{dk}: coefficient of molecular diffusion of species k

In pollutant transport problems, diffusion leads to an omnidirectional propagation of contaminant plumes. However, the influence of diffusion on entire contaminant expansion is generally considered to be negligible due to the domination of advection-based processes in groundwater. Only in few cases (e.g. clayey soils with nearly no fluid motion) diffusion plays a major role in pollutant transport (LaBolle and Fogg, 2001). Diffusion parameters determined in soil column experiments range from $2.5 \cdot 10^{-9}$ m²s⁻¹ (Biswas et al., 1991) to $2 \cdot 10^{-8}$ m²s⁻¹ (Young and Ball, 1998), or $9 \cdot 10^{-8}$ m²s⁻¹ (Chang et al., 2006).

2.2.4 Adsorption

Another crucial process involved in transport modelling is retardation or immobilisation of pollutant molecules caused by adsorption. During adsorption, molecules reversibly adhere to soil particles due to molecular interaction such as van-der-Waals forces or dipole-dipole interaction (Atkins et al., 2001).

Soil organic content as well as hydrophilic properties of the pollutants (e.g. octanol-water distribution coefficient) affect the retention behaviour of substances in the subsurface (Ruffino and Zanetti, 2009). The adsorption is measured experimentally and described mathematically by adsorption isotherms. These adsorption isotherms serve as theoretical tool to calculate the amount of molecules adsorbed onto soil particles, i.e. the Henry-, Freundlich- and Langmuir-concept (Schwarzenbach et al., 2003).

The simplest and most frequently used approach in groundwater modelling is defined according to the Henry-equation (equation (2.6)). In this approach sorption onto soil particles is described by a linear relationship of the concentration of the compound in solution (Atkins et al., 2001). The difference between the Henry-concept and the more sophisticated concepts by Freundlich and Langmuir is in the range of low concentrations (as they occur for groundwater contaminations) and can often be neglected (Ball and Roberts, 1991).

$$L_k = K_{D,k} \cdot C_{k,eq} \qquad (2.6)$$

L = loading of sorbent

$K_{D,k}$ = partition coefficient of species k

$C_{k,eq}$ = solute concentration of species k

A common way to calculate the partition coefficient K_D is based on the octanol-water partition coefficient K_{OW} of the compound of concern, which is converted into K_D using the empirically derived equation (2.7). Many parameter estimations calculated from experimental data have been reported in literature (Schwarzenbach and Westall, 1981; Briggs, 1981; Valsaraj et al., 1999; Karickhoff et al., 1979). All of these references relate contaminant adsorption to the fraction of soil organic carbon, i.e. only the organic carbon fraction (f_{OC}) is assumed to be responsible for adsorption of pollutant molecules. The assumption is made that interaction is only possible between hydrophobic contaminants and hydrophobic organic soil molecules rather than the predominantly hydrophilic bulk soil phase.

The typical form of empirical derivations is given in the following equation:

$$K_D = a \cdot f_{OC} \cdot K_{OW}^{b} \tag{2.7}$$

f_{OC}: fraction of organic carbon

K_{OW}: octanol-water partition coefficient

a, b: experimentally derived parameters

A short overview of the most common equation is given in Table 2.3. Typical values range from 0.3–30 for parameter a and 0.5–1 for parameter b (Ball and Roberts, 1991). As well as the coefficients in equation (2.7), K_{OW} values are determined experimentally. Values can be found in the literature and several databases (e.g. Sangster Research Laboratories, May 4th, 2011) (Table 2.4). Organic carbon fraction is a soil specific parameter which has to be measured for each soil layer separately.

Table 2.3: Empirical equations for calculation of soil partition coefficients K_D according to different authors

1.	$K_D = 6.46 \cdot f_{OC} \cdot K_{OW}^{0.56}$	(Valsaraj et al., 1999)
2.	$K_D = 3.09 \cdot f_{OC} \cdot K_{OW}^{0.72}$	(Schwarzenbach and Westall, 1981)
3.	from $K_D = 4.37 \cdot f_{OC} \cdot K_{OW}^{0.52}$ up to $K_D = 9.55 \cdot f_{OC} \cdot K_{OW}^{0.53}$	(Briggs, 1981)
4.	$K_D = 0.63 \cdot f_{OC} \cdot K_{OW}$	(Karickhoff et al., 1979)

Table 2.4: Octanol-water partition coefficients for different chloroethene compounds

Compound	log(K_{OW}) range [-]	Preferred log(K_{OW}) value* [-]
PCE	2.53 – 3.78	3.40
TCE	2.29 – 2.84	2.42
cis-DCE	1.85 – 1.86	1.86
VC	1.52 – 2.79	1.52

* preferred value according to Sangster Research Laboratories (May 4th, 2011)

If the adsorption coefficient is known, the retention factor R_k for each species is computable (equation (2.8)):

$$R_k = 1 + \frac{\rho_s}{\varepsilon} \cdot K_{D,k} \qquad (2.8)$$

ρ_s = soil dry density

ε = porosity

It is obvious from the explanations given in this section, that implementing adsorption parameters for groundwater contaminant transport models is to a large extent based on empirically derived parameters and that uncertainty for each input parameter ranges in the order of magnitudes. The careful evaluation of adsorption parameter ranges for specific case study areas is hence part and parcel of hydrogeological modelling.

2.2.5 Degradation

In current literature, mainly two different approaches for description of degradation kinetics can be found. The first approach uses constant reaction rates for the different dehalogenation steps of chloroethenes based on a first-order approach, where the degradation kinetic is only dependent on the concentration of the educts (equation (2.9)). If referring to chlorinated ethene decay the terms reaction rate and degradation rate are used synonymously in this work.

$$v_k = -\frac{dC_k}{dt} = -k_k \cdot C_k \qquad (2.9)$$

v_k = reaction rate / degradation rate of compound k

k_k = reaction rate constant for compound k

Most groundwater models, which use reactive transport equations to calculate pollutant expansion on local and regional scale, assume first-order degradation kinetics (Schaerlaekens et al., 1999; Ling and Rifai, 2007). For small contaminant concentrations this approach is non-critical, as non-linear approaches can be approximated by a linear slope in these concentration ranges (Clement et al., 2002). For comparison a list of reaction rates is given in Table 2.1.

Table 2.5: Short overview over published first-order degradation rates for chlorinated ethenes [y^{-1}]

PCE	TCE	cis-DCE	VC	Source
1.83	1.83	1.83	2.19	(Clement et al., 2002)
	30.66	13.51	16.79	(Noell, 2009)
4.02	0.99	1.61	0.58	(Wiedemeier et al., 1998)
	0.40			(Wilson et al., 2001)
2.56	3.29	0.11	0.18	(Ling and Rifai, 2007)
1.10	1.20	1.20	1.72	(Aziz et al., 2000)
1.83	1.83	1.83	1.83	(Schaerlaekens et al., 1999)

Wiedemeier et al. (1998) further collected available degradation rate constants from current literature and analysed their deviations. Deviations ranging from 3-4 orders of magnitude are found (Figure 2.5). Further, mean values of almost 1 y^{-1} (red column) are reported.

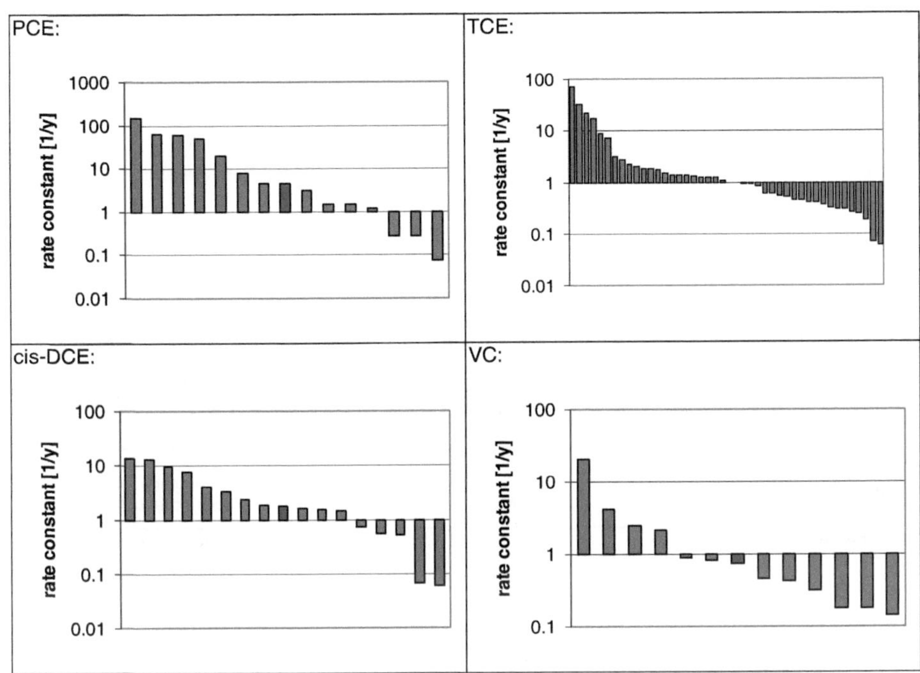

Figure 2.5: Distribution of degradation rate constants (first-order) for chlorinated ethenes, redrawn after (Wiedemeier et al., 1998) for PCE, TCE, VC; own literature assessment for cis-DCE

Another approach to calculate pollutant decay is based on the reaction kinetic described by Monod (1949), which is derived from enzyme kinetic (Michaelis and Menten, 1913). The degradation rates are calculated according to the equation:

$$v_k = -\frac{dC_k}{dt} = -v_{max,k} \cdot \frac{C_k}{K_{S,k} + C_k} \qquad (2.10)$$

$v_{max,k}$ = maximum specific reaction rate for species k

$K_{S,k}$ = half-saturation constant for species k

While the first-order approach is simple and thus often implemented in groundwater modelling, the non-linear Monod approach is more suitable for reactions catalysed by microorganisms, but mainly applied for laboratory scale degradation modelling. In groundwater modelling the more accurate Monod approach is often difficult to apply due to field-data uncertainties which arise from the heterogeneity of the soil-groundwater system and the often low availability of concentration data. Table 2.1 summarises some of the kinetic parameters, which were used for the Monod-inspired kinetic calculations in this work.

Table 2.6: Monod parameters for chlorinated ethene degradation derived from literature

Source	μ_{max}	K_S	Parameter derivation
(Lee et al., 2004)	PCE: 366 TCE: 366 DCE: 48 VC: 48 [µmol/mg/d]	PCE: 0,2 TCE: 0,05 DCE: 3,3 VC: 2,6 [µM]	Batch and CSTR experiments; mixed cultures obtained from contaminated aquifer
(Becker, 2006)	PCE: 43,2 TCE: 72 cDCE: 72 VC: 72 [µmol/mg VSS/d]	PCE, TCE, cDCE: 0,54 VC: 290 [µM]	CSTR experiments; dehalorespiring bacteria like *Dehalococcoides ethenogenes*
(Cupples et al., 2004)		DCE: 3,3 VC: 2,6 [µM]	Batch experiments; dehalorespiring bacteria like *Dehalococcoides ethenogenes*
(Haston and McCarty, 1999)	PCE: 2,0 TCE: 1,6 cDCE: 0,37 VC: 0,34 [µmol/mg VSS/d]	PCE: 0,11 TCE: 1,4 cDCE: 3,3 VC: 2,6 [µM]	Batch experiments; mixed cultures obtained from contaminated aquifer
(Jang and Aral, 2007)	PCE: 0,01 TCE: 0,008 cDCE: 0,0019 VC: 0,0017 [µM/d]	PCE: 0,11 TCE: 1,4 cDCE: 3,3 VC: 2,6 [µM]	Batch experiments; mixed cultures obtained from contaminated aquifer
(Chang and Alvarez-Cohen, 1996)	TCE: 73 cDCE: 86 VC: 89 [µmol/mg/d]	TCE: 47 cDCE: 31 VC: 56 [µM/l]	Batch and CSTR experiments; mixed and pure cultures of methane-oxidizing bacteria
(Amos et al., 2007)	PCE: 129-234 TCE: 259-334 [µmol/mg/d]	PCE: ~14 TCE: 15-23 [µM]	Batch experiments; pure cultures of dehalogenating bacteria
(Haws et al., 2006)	TCE: 2-8 [µmol/mg/d]	TCE: 10-30 [µM/l]	Model development using approximated values for fast and slow degradation scenarios

Stoichiometric factors involved in degradation of chlorinated ethenes can be found in the appendix (section 11.3).

2.2.6   Combined differential transport equations

For modelling purposes the equations for transport and reaction processes (equations (2.3)-(2.10)) can be summarised and merged in one combined differential equation. So, the following formulation is used to calculate reactive transport in porous media is exemplarily formulated as follows (Diersch, 2006):

$$\frac{\partial}{\partial t}\left(\varepsilon_f R_k C_k\right)-\nabla\cdot\left(D_{d,k}\cdot\nabla C_k\right)+\nabla\cdot\left(qC_k\right)-Q_k^{bulk}=\varepsilon_f\left[\left(R_{k-1}+\varphi_{k-1}\right)\upsilon_{k-1}C_{k-1}-\left(R_k+\varphi_k\right)\upsilon_k C_k\right] \qquad (2.11)$$

k = species indicator, $k = 1,...,n$

C_k = concentration of species k

$D_{d,k}$ = tensor of hydrodynamic dispersion

q = Darcy flux

Q_k = zero-order non-reactive production term

υ_k = decay rate

ε = porosity

R_k = retardation factor

Equation (2.11) describes a generalised mass balance for each chemically reacting species in the fluid phase. For solid phase reactions, fluid phase associated transport processes (e.g. diffusion and dispersion) are neglected and the equation simplifies due to omission of the respective terms.

2.2.7 Boundary conditions

Boundary conditions (BC) describe the system's behaviour at the boundaries of the respective domain. Several types of boundary conditions are commonly used, e.g. the Dirichlet BC (1st-kind BC) or the Cauchy BC (3rd-kind BC). With regard to groundwater modelling, boundary conditions are necessary when defining for instance steady-state or dynamic conditions. In the Finite Element (FE) model formulation using Feflow 5.4 as modelling tool, Dirichlet boundary conditions are utilised for defining constant (e.g. groundwater head for steady-state flow conditions) or time-dependant conditions (e.g. mass BCs for contaminant release or head BCs for instationary flow models). The respective BC on each boundary node of the mesh is thus described by a known function.

For example in case of an ordinary differential equation:

$$f\left(x, y(x), y'(x), y''(x)\right)=0 \qquad (2.12)$$

on the interval (a,b) the Dirichlet BC is formulated as follows:

$$y(a)=\alpha, \; y(b)=\beta \qquad (2.13)$$

Another important BC is the 3rd-kind BC or Cauchy BC, respectively. Its main purpose in the model is meant for definition of a transfer rate between groundwater and surface water, which is not only defined by the potential (i.e. the difference in head between both systems), but also by the properties of the interface between them (e.g. grain size distribution of the colmation layer). Thus, it is not only important to know the particular solution of boundary value, but also the value of its derivative.

As an example for the given ordinary differential equation:

$$f(x, y(x), y'(x), y''(x)) = 0 \tag{2.14}$$

on the interval (a,b) the Cauchy BC on one side of this interval can take the form:

$$y(a) = \alpha, \quad y'(a) = \beta \tag{2.15}$$

2.3 Risk assessment approach for groundwater contaminations

2.3.1 Classification of brownfield sites

The high number of registered brownfield sites found in industrialised countries (data for Germany: see appendix, Table 11.9) led to the problem of grading the impact of these sites onto affected environment and population. Hence, risk assessment and management has been developed throughout several disciplines and for different applications such as finance, insurance industry and civil engineering. When considering groundwater contaminations different risk assessment approaches are imaginable, but generally a standardised procedure for risk evaluation during survey of suspicious brownfield sites is not at hand and thus, sophisticated risk assessment for contaminated areas is mostly omitted. Although there are different approaches available (Kaufman et al., 2009; Benekos et al., 2007), risk analysis is performed seldom due to uncertainties in contaminant transport and elements at exposure. The risk management framework developed by Pliefke et al. (2007) combines different aspects of risk management in a holistic framework (Figure 2.6).

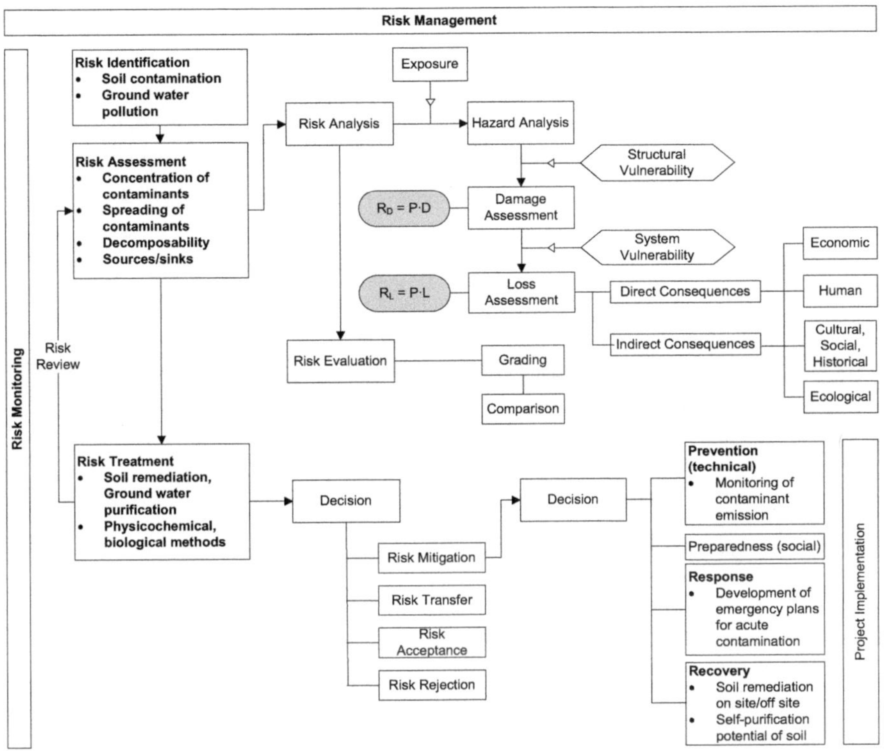

Figure 2.6: Proposed probabilistic risk management framework for soil contaminations, redrawn and extended after Pliefke et al. (2007)

A somewhat similar treatment is intended to be performed by German authorities in order to deal with soil pollution and to guarantee public safety: the Bundes-Bodenschutzgesetz (BBodSchG, 1998) as well as the Bundes-Bodenschutzverordnung (BBodSchV, 1999) propose a multi-stage advance for risk identification, assessment and treatment of brownfield sites.

In the first step (step 1: "survey without sampling", risk identification phase) an identification of potential soil contaminations is performed, e.g. by analysis of aerial photographs, file inquiry of former industry companies or on-site inspection. This procedure is initiated if any evidence for possible contaminations exists and, in case of a positive outcome, leads to documentation of the suspicious area in a so-called brownfield register.

The second step (risk assessment phase) is divided into two sub-steps. In the first one (step 2a: "preliminary examination") soil and groundwater samples are collected in a wide-meshed grid and the detected contaminants concentrations are compared with corresponding inspection values (published in BBodSchV). The exceedance of these inspection values leads to the second step (step 2b: "detailed examination"), which considers the horizontal and vertical extent of the soil contamination is investigated. In case of no or only a punctual exceedance of threshold concentrations, the suspicion of pollution is dropped. However, in the opposite case a hazard analysis needs to be performed by the responsible environmental authority in order to evaluate the potential risk arising from the pollutants for the environment in general and nearby residents in particular. Decisions during this part of risk assessment are normally made on the basis of land and groundwater use as well as on expected contaminant fate during future decades. Thus, it is crucial to have fundamental estimations about the contamination development in the area, which is best acquired by groundwater models.

The result of this decision process is classification of the affected area as brownfield site, leading to further studies about feasibility of remediation measures (step 3) and grading of the expected costs and benefits for different alternative clean-up activities (risk treatment phase).

2.3.2   Risk management concepts

When dealing with risk assessment, some basic considerations need to be performed concerning the terminology in order to avoid ambiguity. Typically different elements at risk (e.g. houses, persons) with different vulnerabilities (masonry vs. wooden houses, young vs. old persons) are exposed to a certain hazard (flood, toxic substances) that occurs with a given probability (in events/year, ...) and leads to a particular damage or loss (monetary or life loss) (Pliefke et al., 2007).

Using these preliminary considerations a mathematical description of risk is possible, given by the following equation:

$$Ri = P \cdot D \tag{2.16}$$

Ri = risk

P = probability of occurrence of a fatal event

D = damage/consequences of this event

The advantages of this formula become evident, if a certain probability for a catastrophe can be derived from historical data (e.g. flood or earthquake occurrence) and if the structural damage (e.g. loss of houses, infrastructure) is quantifiable, i.e. the elements at risk do have a monetary value. It becomes more intricate when economic and cultural losses or even losses of human life are taken into account.

Other problems associated with equation (2.16) are mainly caused by either uncertainty in probability or quantification of consequences. Thus, sophisticated methods based on statistical analysis have been developed in the past in order to deal with these problems (first-order reliability method (FORM), six-sigma-design etc.). Nevertheless, in the field of environmental risk assessment as discussed in this work, this method does not appear to be feasible due to the fact that environmental consequences of groundwater pollutions can hardly be described in terms of monetary damage. However, a risk treatment approach can be derived from a modification of equation (2.16) if grading remediation options by minimisation of costs. The objective of comparing clean-up techniques is basically a cost-benefit analysis, where the probability of a fatal event (in this case P corresponds to the consequences of soil pollution) is reduced to zero by eliminating the pollution itself. Thus, the damage term (D) can be seen in monetary aspects (i.e. as remediation costs for complete contaminant removal) which are subject to minimisation in order to gain the largest benefit out of the least costs. Such approaches regarding optimisation of remediation costs can be found in literature (Stupp et al., 2005; Kaufman et al., 2005).

While the approach described above primarily deals with treatment of risk (according to Figure 2.6), another goal of this work is the assessment of risk caused by groundwater contaminations. The key issue arising from this advance is the matter of potential risk receptors, or in different words: who or what is affected by the pollution and does it have negative consequences for these elements at risk?

Given the fact that contaminated groundwater might be used for household purposes by people living in the affected area, it becomes obvious that human health may be of concern for risk assessment studies. Additional pathways of exposure towards chemical substances in soil are conceivable according to the German Bundes-Bodenschutzverordnung:

- Soil → groundwater pathway: usage of contaminated groundwater for drinking, showering or cooking
- Soil → agricultural crop pathway: incorporation of pollutants into crop roots and successive ingestion as food
- Soil → human pathway: exposure to contaminated soil via lung (dust in the air), dermal (contact with soil) or gastrointestinal contact (uptake of soil)

The calculation of exposure via each of these pathways is explained in detail in Ma (2002).

While the soil-crop pathway is of special concern in regions with agricultural land use, contact with soil is supposed to occur predominantly at children playgrounds and areas with high recreational usage (e.g. soccer courts, sunbathing lawn). The soil-groundwater pathway is of particular importance if groundwater is used for household purposes. Contact with and uptake by humans is imaginable via ingestion of drinking water, ingestion of shower water, inhalation of shower air and dermal contact of shower water. Different simulation programs have been developed in the past to estimate the exposure of chemical substances and the inherent risk via the respective pathways. The most established tools for multimedia risk assessments are 3MRA (Marin et al., 2003), MEPAS (Whelan et al., 1992), MMSOILS (US EPA, 1997) and CalTOX (McKone, 1993). Several studies on comparison of these programs (Mills et al., 1997; Chen and Ma, 2006) as well as on the impact of different pathways (Fan et al., 2009; Piver et al., 1997) can be found in literature. All of the above mentioned studies indicate that ingestion of drinking water contributes to an extraordinary extent to the overall hazard posed by groundwater contaminations. For this reason authors tend to focus on oral ingestion of pollutants (Benekos et al., 2007; Massabo et al., 2008) and neglect the remaining pathways, which additionally are difficult to calculate and as well comprise a high amount of uncertainty. This simplification was also adopted for the basic concept of the risk assessment approach developed in this work.

2.3.3 Health risk assessment approaches

In the early 1980ies concerns about the environmental impact of chloroethenes and the influence on human health rose. Several studies regarding the effects of these chemicals were conducted and summarised in toxicological profile reports, published by the U.S. Agency of Toxic Substances and Disease Registry (ATSDR) of the U.S. Department of Health and Human Services, or in toxicological reviews by the U.S. Environmental Protection Agency (U.S. EPA). Currently, updated reports for all chlorinated ethenes are available from both, ATSDR (http://www.atsdr.cdc.gov/) and U.S. EPA (http://www.epa.gov/).

The recommended cancer risk assessment approach performed by these agencies is based on calculating the potential cancer risk posed by carcinogenic substances for persons concerned. The cumulative cancer risk is then assumed as the sum of cancer risks via different pathways (as described in chapter 2.3.2). An exposure rate which is defined as average daily dose (ADD) is computed for each pathway and multiplied with a compound-specific cancer slope factor (CSF). Cancer slope factors, determined by different animal experiments and bioassays, can also be found in the respective literature of both, ATSDR and U.S. EPA (Table 2.7). Three of the four different chlorinated ethenes discussed in this work have proven to possess carcinogenic properties (PCE, TCE, VC) and have therefore been subject to cancer risk evaluation.

The total cancer risk can be obtained by summation of the risk for each substance (Maxwell and Kastenberg, 1999; Piver et al., 1997). The mathematical relation is given by the following equation:

$$Ri = \sum_{j=1}^{m} \sum_{k=1}^{n} ADD_{k,j} \cdot CSF_k \qquad (2.17)$$

k: species indicator (PCE, TCE, DCE, VC)

j: index of different pathways

ADD: average daily dose [mg·kg^{-1}·d^{-1}]

CSF: cancer slope factor [(mg·kg^{-1}·d^{-1})$^{-1}$]

The carcinogenicity assessment provides information on the carcinogenic hazard potential of the substance in question and quantitative estimates of risk from oral exposure. The information includes a weight-of-evidence judgment of the likelihood that the agent is a human carcinogen and the conditions under which the carcinogenic effects may be expressed. Quantitative risk estimates may be derived from the application of a low-dose extrapolation procedure. If derived, the cancer slope factor is an upper bound on the estimate of risk per mg·kg^{-1}·d^{-1} of oral exposure (U.S. EPA). When assuming that oral ingestion of contaminated groundwater accounts for the major part of the total risk, the average daily dose is described in the following way (McKone, 1991):

$$ADD_k = \overline{C}_{kt} \left(\frac{GR}{BW} \right) \cdot \frac{ED \cdot EF}{AT} \qquad (2.18)$$

$\overline{C}_{kt}$: average concentration of contaminant k in tap water over an exposure time t [mg·l^{-1}]

GR: ingestion rate of tap water [l·d^{-1}]

BW: body weight [kg]

ED: exposure duration [y]

EF: exposure frequency [d·y^{-1}]

AT: averaging time [d]

This approach has been performed and applied to different groundwater pollution scenarios by several authors (Benekos et al., 2007; Massabo et al., 2008; Lemke and Bahrou, 2009) based on the assumption, that oral ingestion of groundwater as drinking water is the main exposure pathway for chlorinated ethenes. Given the fact that the CSF includes a safety factor to avoid underestimation of the resulting risk, the major problem consists of calculation of the average daily dose. Uncertainties concerning this exposure rate remain, primarily due to unknown values of total concentration of pollutants in the drinking water and the heterogeneity of the affected population (i.e. the vulnerability). It is therefore of major importance to have precise estimations of future concentration development at affected sites, which is best obtained by accurate and calibrated groundwater models. Benekos et al. (2007) further developed a probabilistic approach to assess uncertainty of the entire cancer risk prediction chain using a two-stage Monte-Carlo (MC) simulation advance based on the software 3MRA. This calculation also includes uncertainty assessment of affected residents, e.g. arising from different body weight or exposure duration, delivering a sophisticated review of uncertainty connected with variation in population characteristics. However, their calculations are performed only for an artificial block-shaped aquifer model without variability in structural parameters.

While the considerations described above concern risk assessment of carcinogenic substances a different approach has to be chosen in order assess the health impact of toxic, but not carcinogenic chemicals. The advance proposed by the U.S. EPA in this case compares the ADD of a certain substance with a respective daily dose that is regarded as safe. This dose is defined as reference dose (RfD) for the chemical of concern. The RfD provides quantitative information for use in risk assessments for health effects known or assumed to be produced through a nonlinear mode of action. The RfD (expressed in units of $mg \cdot kg^{-1} \cdot d^{-1}$) is defined as an estimate (with uncertainty spanning of perhaps an order of magnitude) of a daily exposure to the human population that is likely to be without an appreciable risk of harmful effects during a lifetime. Reference values are generally derived for chronic exposures (up to a lifetime), but may also be derived for acute (24 hrs), short-term (>24 hrs up to 30 days), and subchronic (>30 days up to 10% of lifetime) exposure durations, all of which are derived based on an assumption of continuous exposure throughout the duration specified (U.S. EPA).

The mathematical expression for calculation of the Hazard Quotient (HQ) is given in the following equation which delivers estimations, if human exposure towards a toxic substance is potentially harmful (HQ > 1) or probably harmless (HQ < 1):

$$HQ = \sum_{j=1}^{m} \sum_{k=1}^{n} \frac{ADD_{k,j}}{RfD_k} \qquad (2.19)$$

HQ: Hazard quotient [$mg \cdot kg^{-1} \cdot d^{-1}$]
RfD: reference dose [$mg \cdot kg^{-1} \cdot d^{-1}$]

Validated values for both, cancer slope factors CSF and reference doses RfD are published in the toxicological reviews of the U.S. EPA. These values were derived from various animal experiments and bioassays and comprise an uncertainty of around one order of magnitude. In all cases several safety factors were applied, e.g. a factor of 3 for the scaling effect from animal experiment to human exposure.

Table 2.7: Summary of cancer slope factors (CSF) and reference doses (RfD) for chlorinated ethenes

Compound	Cancer slope factor [$mg \cdot kg^{-1} \cdot d^{-1}]^{-1}$	Reference dose [$mg \cdot kg^{-1} \cdot d^{-1}$]	Source
PCE	0.01 – 0.1	0.004	(U.S. Environmental Protection Agency, 2008)
TCE	0.05	0.0004	(U.S. Environmental Protection Agency, 2009)
cis-DCE	---	0.002	(U.S. Environmental Protection Agency, 2010)
VC	0.72	0.003	(U.S. Environmental Protection Agency, 2000)

Draft versions of the reviews for PCE and TCE were chosen in the table above due to the high age of former review versions (1988 for PCE and 1999 for TCE, respectively). Previous reports on toxicity of chloroethenes provided by the U.S. EPA generally stated higher values for the reference dose (e.g. for PCE: 0.01 $mg \cdot kg^{-1} \cdot d^{-1}$ in the 1988's report in comparison to 0.004 $mg \cdot kg^{-1} \cdot d^{-1}$ in the preliminary report from 2008) and similar values for the cancer slope factor (e.g. 0.0511 $mg \cdot kg^{-1} \cdot d^{-1}$ in the report of 1988 and 0.01 – 0.1 $mg \cdot kg^{-1} \cdot d^{-1}$ in the current report) (U.S. Environmental Protection Agency, 1988, 2008)

For the assessment of health risks posed by groundwater contaminations towards residents, the ADD is computed using the different transport and reaction models described in chapter 5. The calculation of two different health risk approaches is performed according to equations (2.17) and (2.19). Additionally, an uncertainty evaluation for the calculated contaminant concentrations is conducted and the results are described in chapter 6 using the groundwater models developed in chapter 5.

3 Experimental area

3.1 *Geological setting and contaminant release*

Field experiments and data acquisition refer to an area of nearly 0.4 km² that is located in Braunschweig, Germany (Figure 3.1). This area was chosen as reference area for a risk assessment approach on the field scale and to simulate chemical and hydrogeological processes in different environmental scenarios.

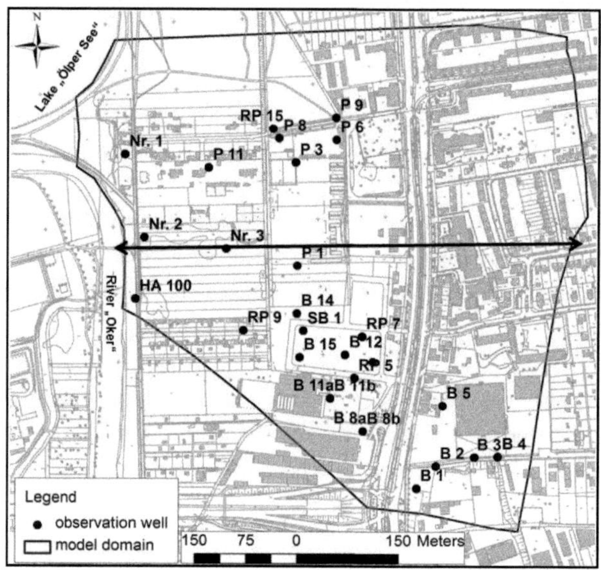

Figure 3.1: Investigation area with model domain and observation wells; horizontal line represents a petrographic cross-section shown in Figure 5.2

The sedimentary aquifer in the study area comprises a thickness of 20-40 m and is divided in around 10 m depth by a less permeable silt layer into an upper and lower part. The catchment is characterised partially by an impervious surface due to buildings and roads (mainly in the eastern part of the domain) and additionally by some natural areas (mostly located in the western area). The whole municipal area consists of cretaceous sediments under Holocene and Pleistocene accumulations. Fluviatile sediments predominantly occur in the western part near the lake and river (Stegmann, 1969). Hydraulic conductivity was determined by Slug tests and ranges from about

1.2·10^{-4} to 3·10^{-3} m·s^{-1}. Analyses of the grain size distribution from several bore drillings and subsequent comparison with literature confirm these values. Different soil layer properties according to these evaluations are given in Table 3.1, an exemplary overview of the soil stratigraphy is depicted in Figure 5.2.

Table 3.1: Soil layers present in the case study area (soil type by analysis of drilling profiles, hydraulic conductivity derived from literature, organic carbon fraction determined by own measurements)

Layer	Soil type	Hydraulic conductivity [10^{-4} m·s^{-1}]	Organic carbon fraction [%]
1	Rubble	5	0.1
2	Silt	1.16	2.5
3	Fine sand	4.63	0.8
4	Middle sand	23.15	0.2
5	Fine sand	4.63	0.2
6	Coarse sand	30	0.1
7	Middle sand	23.15	0.2
8	Fine sand	4.63	0.2
9	Coarse sand	30	0.1
10	Silt	1.16	2.5

Groundwater flow in the area is directed from the south-east to the north-west. Given the above mentioned hydraulic conductivities and a gradient of around 1‰, flow velocities of approximately 36 m·y^{-1} can be deducted (calculations by the engineering office "Geolog").

In the south-eastern part of the domain, a former chemical laundry disposed their sewage water into the soil, leading to contamination predominantly consisting of PCE and its degradation products (TCE, cis-DCE and VC). Since several decades the pollutants have been subject to transport and degradation processes, resulting in a contamination plume of approximately 500 m length and 100 m width nowadays. The major contaminant hotspots can be found in depths up to 10 m, where a silt layer divides the aquifer into a primary and a secondary aquifer. Samples from 25 m depth contained nearly no pollutants. Hence, the secondary aquifer is not taken into account for groundwater modelling, due to the fact that contaminant occurrence is mainly observed in the upper, sandy layers.

Particularly, the long-term contaminant mass fluxes into boundary ecosystems on the western border such as the nearby lake (lake Ölper) and river (river Oker) are of great concern. The leakage of contaminated groundwater into these ecosystems is to be avoided in first place. In addition, not only seen from an environmental point of view but also regarding health aspects, threat is posed towards the approximately 20-30 residents living in the north-western part of the area, who did not possess municipal drinking water-supply until summer 2010 (according to personal communication with residents). In the worst-case scenario it may be assumed that drinking water demand as well as other household water needs were satisfied by utilisation of groundwater. Thus, the health risk approach proposed in chapter 2.3.3 is supposed to be a suitable method in order to assess risk for affected residents as a worst-case scenario.

3.2 Field measurements

In the area of concern several groundwater observation wells are available. In order to validate the numerical model developed in this work, samples from 20 observation wells were taken and analysed for chlorinated ethenes using an Agilent 7890 Headspace GC/MS system and the certified analysis laboratory "Biolab Umweltanalysen GmbH". Additionally, several environmental parameters (pH, redox potential, temperature etc.) were measured and concentrations of different ions like nitrate, sulphate and chloride were analysed with a VWR-Hitachi LaChrom HPLC system (see appendix for materials and operational procedures). The concentrations of the four pollutant species (PCE, TCE, cis-DCE and VC) were also determined over several years of monitoring (1996 to 2010), which result in a comprehensive overview of the current pollution situation. Recently analysed contaminant concentrations exceeded 40.000 $\mu g \cdot l^{-1}$ of total DNAPLs in the core area (reported by the engineering office "geo-log GmbH", 2006). To give an impression about the contamination plume extent, concentration isolines generated via inverse distance interpolation are shown in Figure 3.2 - Figure 3.5. However, it has to be denoted that interpolation does not work faultless and that these isolines are only approximated on the basis of measurement data from around 20 observation wells located in the area (shown in Figure 3.1).

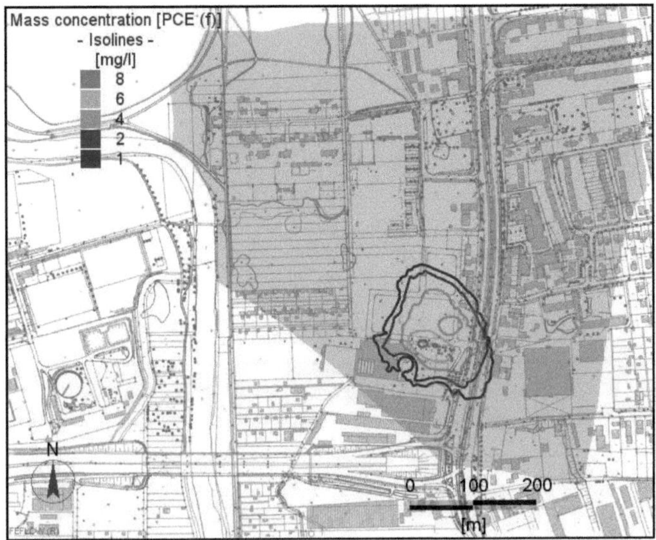

Figure 3.2: Interpolated contaminant plume at the case study area "Schützenplatz" for PCE

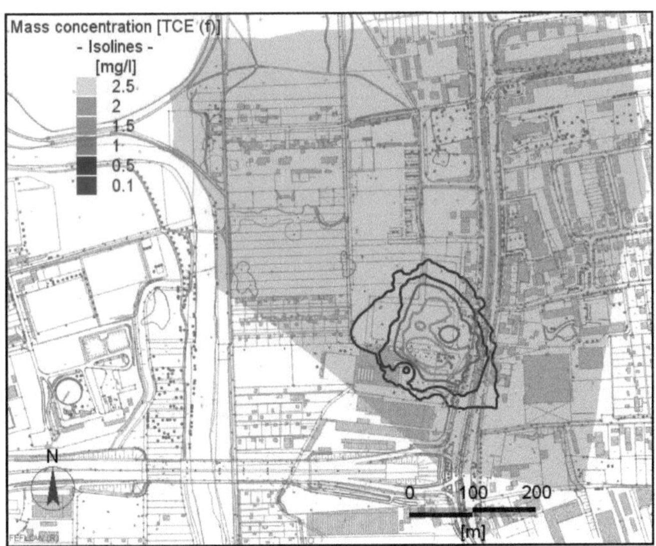

Figure 3.3: Interpolated contaminant plume at the case study area "Schützenplatz" for TCE

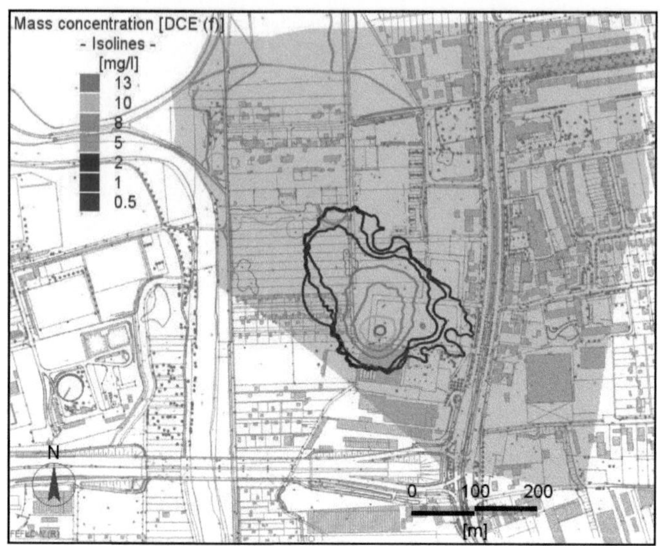

Figure 3.4: Interpolated contaminant plume at the case study area "Schützenplatz" for cis-DCE

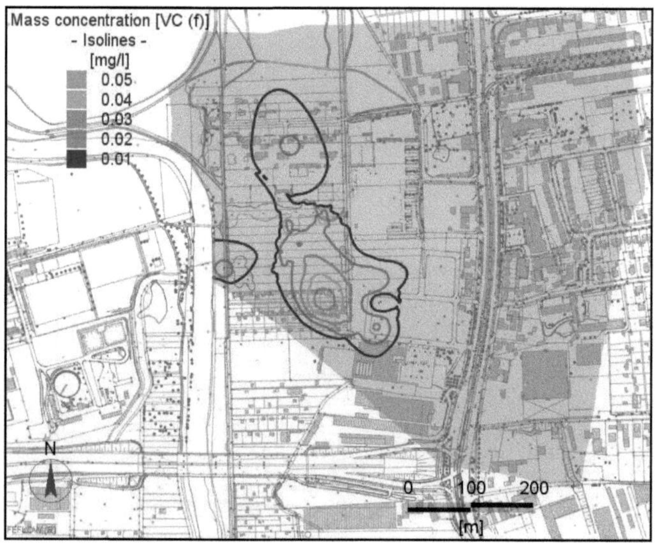

Figure 3.5: Interpolated contaminant plume at the case study area "Schützenplatz" for VC

During measurement campaigns the concentrations of the most important inorganic electron acceptors were analysed using HPLC- and AAS-methods (as described in the appendix, chapter 11.6). Those substances were dissolved oxygen, nitrate, iron, manganese and sulphate. Given the fact, that iron(III) and manganese(IV) have a very low solubility constants in water, all measured iron and manganese was considered to be in the reduced form (iron(II) and manganese(II), respectively) (Azadpour-Keeley et al., 2001). Table 3.2 exemplarily summarises the measured values of all survey campaigns for some of the groundwater wells (additional data see appendix, Table 11.10a-c).

Table 3.2: mean measurement values and standard deviation for exemplarily chosen observation wells

	B12	SB1	TUBS1
PCE [µg·l⁻¹]	4432 ± 2053	388 ± 284	3 ± 6
TCE [µg·l⁻¹]	1586 ± 894	82 ± 48	0 ± 0
cis-DCE [µg·l⁻¹]	6488 ± 2322	8245 ± 1478	453 ± 356
VC [µg·l⁻¹]	6 ± 8	4 ± 6	5 ± 6
pH value [-]	6.9 ± 0.1	7.0 ± 0.1	6.9 ± 0.1
conductivity [mS·cm⁻¹]	1606 ± 262	1544 ± 136	1436 ± 142
redox potential [mV]	200 ± 54	13 ± 7	-89 ± 21
diss. Oxygen [mg·l⁻¹]	0.8 ± 0.3	0.8 ± 0.4	0.9 ± 0.5
diss. organic carbon (DOC) [mg·l⁻¹]	5 ± 4	4 ± 4	7 ± 2
nitrate [mg·l⁻¹]	11 ± 5	1.5 ± 2.4	0.4 ± 0.7
nitrite [mg·l⁻¹]	0.5 ± 0.6	0.5 ± 0.7	0.5 ± 1.1
chloride [mg·l⁻¹]	98 ± 26	99 ± 21	94 ± 20
sulphate [mg·l⁻¹]	542 ± 113	491 ± 41	460 ± 31
sodium [mg·l⁻¹]	97 ± 10	79 ± 17	59 ± 13
potassium[mg·l⁻¹]	22 ± 6	12 ± 4	12 ± 4
magnesium [mg·l⁻¹]	35 ± 7	32 ± 5	43 ± 5
calcium [mg·l⁻¹]	274 ± 47	239 ± 12	235 ± 39
diss. iron [mg·l⁻¹]	0.2 ± 0.2	0.2 ± 0.1	13.8 ± 1.1
diss. manganese [mg·l⁻¹]	1.9 ± 0.4	1 ± 0.1	2.9 ± 0.5

The results of the analysis revealed, that oxygen is present in all observation wells at low concentrations of almost 1 mg·l⁻¹. However, due to the sampling procedure it is possible that small amounts of oxygen dissolved from the air into the samples during the in-situ measurement (e.g. caused by cavitation in the rotary pump). This assumption is emphasised by the almost constant value of oxygen in all analysed samples with only small deviations. Hence, for the model establishment anaerobic conditions are assumed all over the experimental area.

The sulphate concentrations observed in the experimental domain mainly ranged from 400 - 500 mg·l^{-1} with some outliers below and above these values. However, there was no observable tendency in sulphate measurement data and so, no variation of degradation kinetics caused by spatially changing sulphate concentrations is assumed in the model described below.

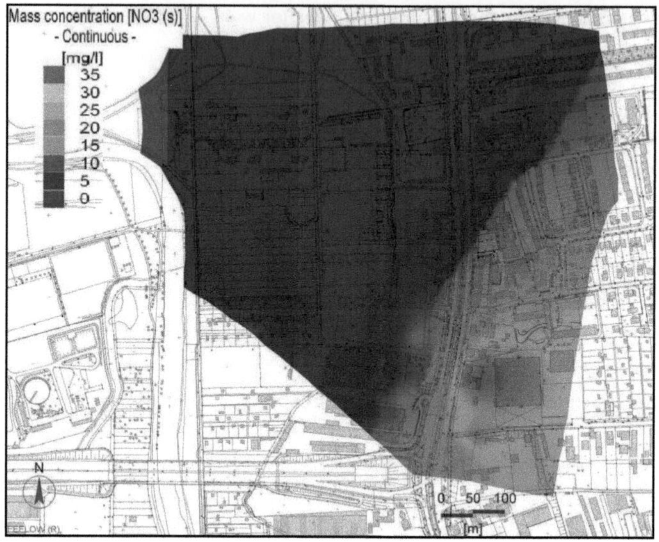

Figure 3.6: Interpolated nitrate concentrations at the Schützenplatz area based on measurement data

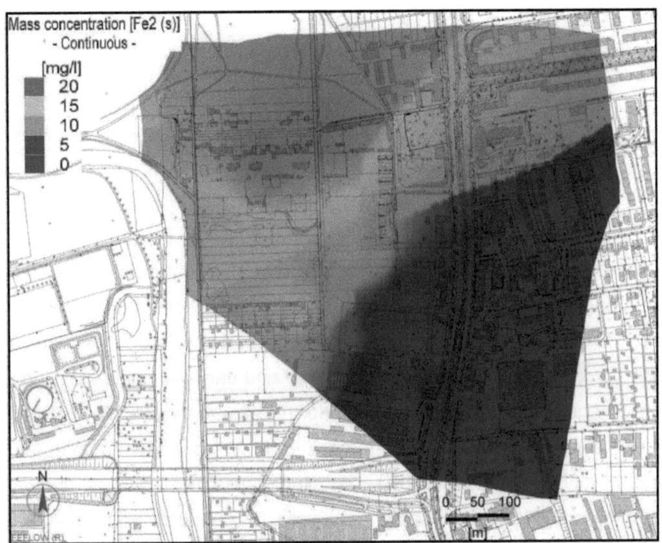

Figure 3.7: Interpolated dissolved iron concentrations at the Schützenplatz area based on measurement data

The occurrence of the remaining electron acceptors nitrate and iron(II) is depicted in Figure 3.6 and Figure 3.7. For manganese(II) a distribution similar to iron(II) was found, but comprising lower total concentrations (of up to 5 mg·l⁻¹). It becomes obvious, that nitrate is only present in the upstream region (eastern part) and drops to zero in a sharp decline, while the concentrations of iron and manganese increase further downstream (western part). This observation is confirmed by redox potential measurements, which show a decline from ~+200 mV in the south-eastern part to ~-100 mV in the north-western part near river Oker and lake Ölper.

The distribution of chlorinated ethenes in the area also exhibits a connection with the occurrence of inorganic electron acceptors: where nitrate is present, PCE and TCE can be found, whereas an increase in iron and manganese concentrations corresponds with a shift in chloroethene spectrum towards cis-DCE. This fact leads to the assumption, that the degradation kinetic of chloroethenes is highly dependant on the redox conditions, represented by nitrate, manganese and iron ion concentrations (Doong et al., 1996; Widdowson, 2004).

4 Laboratory and soil column experiments

4.1 Adsorption and retardation soil column experiment

4.1.1 Materials and methods

Soil column experiments were conducted for evaluation of different transport parameters of chlorinated ethenes. A sandy soil from the case study area was obtained during borehole drilling activities and was used for this experiment in order to apply natural conditions. Soil columns of 70 cm in length and 7.2 cm in diameter were filled with this soil and uncontaminated groundwater was recirculated overnight to equilibrate the columns. During the experiments sodium chloride, which has shown to exhibit no significant sorption onto the soil in preliminary experiments, along with a mixture of chlorinated ethenes was injected into the column bottom as a Dirac impulse. At the upper end of the column samples were taken and analysed for conductivity, pH value and redox potential. Additionally, chlorinated ethene concentrations were determined using an Agilent 7890 Headspace GC/MS system. A flow of about 9 ml·min^{-1} was chosen for all experiments, which were conducted twice including repeat determination, resulting in a total of four concentration datasets. Figure 4.1 depicts the experimental setup for the adsorption tests as well as a photo of one of the soil columns.

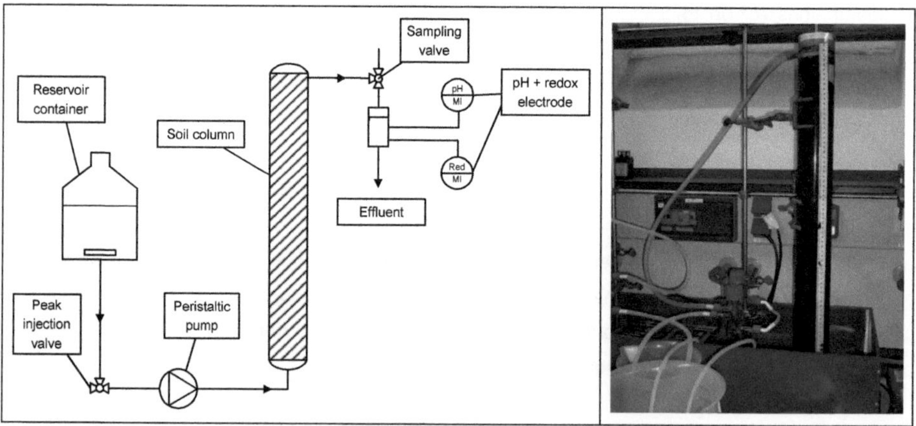

Figure 4.1: Experimental setup and photo of the laboratory soil column

In a further step contaminant transport is tried to be modelled in Feflow using a true-to-scale 2D column model with 249 4-noded quadrilateral elements and implying the soil properties determined in previous determinations (model setup and settings: see appendix, chapter 11.1).

4.1.2   Results of the experiment

The results obtained from the soil column experiments are exemplarily shown in Figure 4.2. For the sake of clearness and readability only averaged values for conductivity, PCE and cis-DCE are depicted, each normalised on the maximum measured or rather simulated concentration of the respective chemical species. Results for the unembodied species TCE and VC show a comparable behaviour (see appendix: Figure 11.1). It becomes obvious that measured and simulated results fit comparatively well and thus, modelling of the soil column appears to be a feasible tool to describe chlorinated ethene transport.

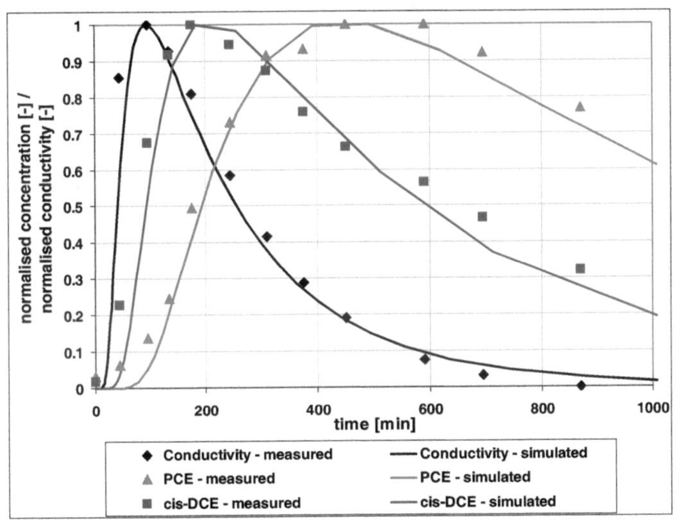

Figure 4.2: **Measured (dotted) and simulated (solid) normalised concentrations /conductivity of the soil column experiments to determine transport behaviour of a conservative tracer (NaCl, shown as conductivity), PCE and cis-DCE**

For modelling purposes diffusion and reaction terms are negligible due to the small time scale of the experiment. Therefore the only parameters left to describe transport are adsorption, porosity and dispersion. Porosity is assumed to possess a constant value of 0.3 throughout the whole column. While transversal dispersion can be neglected due to the experimental setup, longitudinal dispersion is unknown and needs to be estimated. A value of 0.4 m worked well for all column simulations.

In theory adsorption of chlorinated ethenes can be calculated based on the theoretical background discussed in chapter 2.2.4. However, the large range of possible values resulting from the approaches makes it necessary to assess the most probable solution for each compound separately. Table 4.1 summarises the parameters resulting from both, experiment and simulations.

Table 4.1: Sorption parameters determined by experiment and simulation

Compound	Range of retention factors (by experiment) [-]	K_D value (used for simulation) [-]
PCE	3.6 – 4.1	$6.4 \cdot 10^{-3}$
TCE	1.7 – 2.8	$3.6 \cdot 10^{-3}$
cis-DCE	1.6 – 2.0	$1.9 \cdot 10^{-3}$
VC	1.0 – 1.2	$2.8 \cdot 10^{-4}$

In an additional experiment the capability of Feflow to simulate reactive transport processes was evaluated. In this test a certain amount of the heavy metal cadmium was injected into the soil column. The high K_D value of this metal leads to quick adsorption onto the soil particles. After establishment of equilibrium a solution containing the chelating agent EDTA was continuously injected into the column, leading to desorption of cadmium due to formation of thermodynamically preferred, highly water-soluble cadmium-EDTA complexes. The concentrations of both, EDTA (via HPLC) and cadmium (via AAS) were measured at the outlet of the column. Based on theoretical considerations a breakthrough curve of cadmium along with rising concentrations of EDTA is expected in this experiment. This was proven by the obtained experimental data (Figure 4.3).

In the simulation programme this behaviour is realised by a reaction of the components cadmium (high soil adsorption behaviour) and EDTA (low soil adsorption behaviour) which form cadmium-EDTA (low soil adsorption behaviour). According to the fast reaction rate, the application was performed by utilisation of a first-order reaction kinetic with high reaction constants.

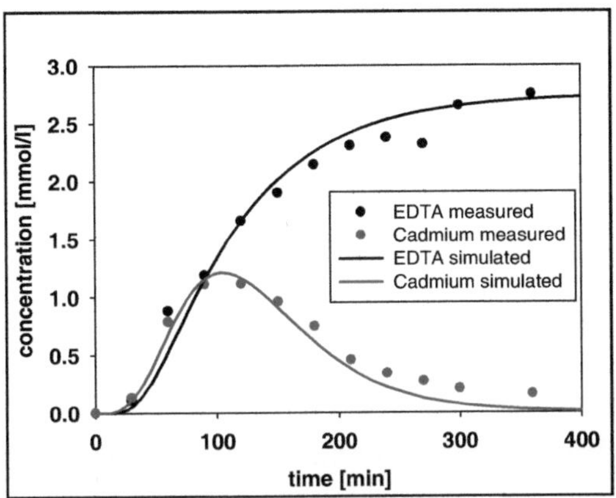

Figure 4.3: Soil column experiment showing heavy metal desorption with EDTA; comparison of experimental and simulation results

The concentration curve is depicted in Figure 4.3 and reveals good agreement of measured and simulated curves. Even the correct molar ratios were represented by the simulation, leading to the conclusion that it is possible to model reactive transport even with more complex problems like the sequential chlorinated ethene degradation chain.

4.2 Batch experiment of chlorinated ethene degradation

4.2.1 Materials and methods

The experiment described in this chapter was conducted for assessment of degradation rates of chlorinated ethenes under laboratory conditions and to find possible substrates for an in-situ remediation on the field scale using addition of degradation-enhancing substances. Additionally, the effects of different inorganic electron acceptors onto degradation kinetics were under evaluation. Five different test batch approaches were conducted each using different initial conditions.

- In a reference sample which only contained soil and polluted groundwater from the field-site, the kinetic of unaffected contaminant degradation was tested.
- Two other batch tests were performed with addition of glucose or EHC, respectively. EHC is a mixture of zero-valent iron and an organic compound and, according to the manufacturer

Adventus, was invented especially for remediation of chloroethene groundwater contaminations. Both, glucose and EHC are supposed to serve as nutrients for ubiquitous, autochthonous soil microorganisms leading to an increased growth of potentially contaminant degrading bacteria. As a further effect, glucose and EHC lower the redox potential due to consumption of electron acceptors during their decomposition. In addition, zero-valent iron in the EHC mixture is presumed to catalyse an abiotic dechlorination of chlorinated ethenes.

- In the fourth batch test nitrate is added to soil and groundwater in order to evaluate the inhibiting effect of inorganic electron acceptors onto reductive dechlorination. This inhibition is assumed to occur due to the thermodynamically favoured denitrification reaction in comparison to reductive dechlorination (see chapter 2.1.2).
- An additional Nitrate-EHC batch test was conducted to check the interference of nitrate inhibition during addition with EHC.

An overview of the different initial conditions for all test batches is given in Table 4.2.

Table 4.2: Setup of the batch experiment for assessment of chloroethene degradation

Batch number	Name	Mass soil [g]	Mass EHC [mg]	Mass glucose [mg]	Mass sodium nitrate [mg]
1	Reference sample	50	-	-	-
2	EHC experiment	50	250	-	-
3	Glucose experiment	50	-	250	-
4	Nitrate experiment	50	-	-	34.3
5	EHC-Nitrate experiment	50	250	-	34.3

The experiment was performed in 250 ml wide-necked brown glass jars (DIN 55). The components of each test were added and the jar was filled bubble-free with groundwater from the field-site and tightly sealed with a Teflon-coated lid. The experiment was conducted at room temperature; samples were shaken once a day during the incubation time. All experiments were conducted in double determination. As the sample needs to be sacrificed for determination of chloroethene concentrations, one jar was necessary for each sampling point. Five sampling times were chosen at 4, 11, 18, 29 and 38 days, resulting in a total amount of 50 jars. Directly after opening the respective samples, redox potential, dissolved oxygen and electric conductivity was measured and a small volume of sample was taken for analyses of cation and anion concentrations, DOC (operational procedures as described in the appendix, chapter 11.6) and chlorinated ethenes.

4.2.2 Results of the experiment

The analysis of the concentration data revealed interesting insight into the mechanism of chlorinated ethene degradation. Figure 4.4 shows the chloroethene concentrations for the experiments 1 – 4 (reference sample, EHC sample, glucose sample, nitrate sample). The EHC-nitrate experiment (experiment 5) basically exhibited similar behaviour to the experiment with EHC only (Figure 4.4b) but with a certain delay of PCE degradation at the start of the experiment, during which nitrate is completely consumed.

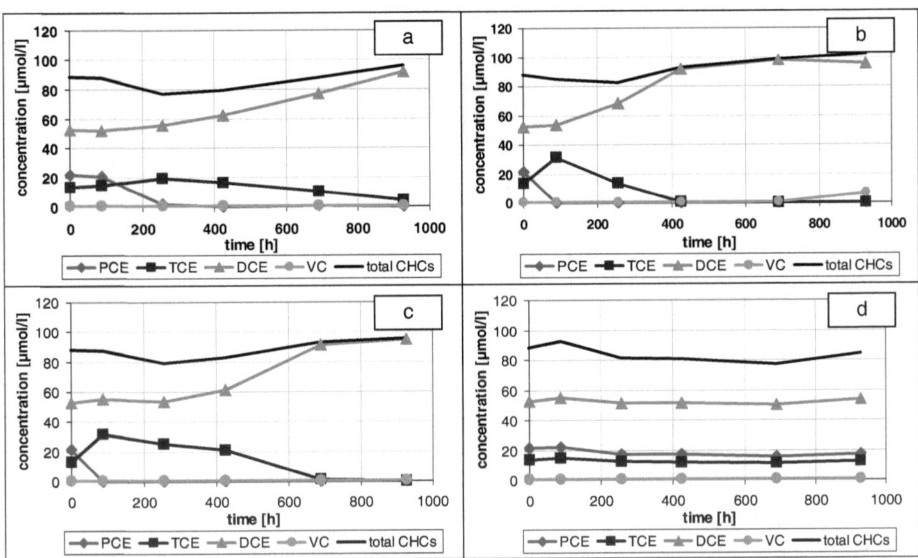

Figure 4.4: Chlorinated ethene concentrations in the different batch tests: a) reference sample; b) EHC sample; c) glucose sample; d) nitrate sample

The experimental data plotted in Figure 4.4 and Figure 4.5 depicts the mean concentration data of both the double determinations. The reproducibility of the experiment was extraordinarily high and the deviation from the mean was generally too small to depict recognisable error bars. Additionally, it could be shown that the total concentration of all chloroethenes remains constant over the whole experimental time. Hence, the general design of the experiment appears to be a feasible method for assessment of the degradation potential and to test different soil remediation strategies.

It becomes obvious from Figure 4.4d that the experiment containing nitrate shows almost no change in contaminant concentrations. Along with Figure 4.5d it can be deducted that nitrate reduction is preferred over the degradation of PCE and thus, the degradation chain of chlorinated ethenes is completely inhibited until nitrate is depleted. The data for the EHC-nitrate experiment also confirm this observation: PCE starts being degraded after a quick decline of nitrate concentration.

In the samples without nitrate addition, this substance is only present in small amounts and hence does not play a crucial role for inhibition of reductive dechlorination. However, even these small concentrations of almost 0.03 mmol·l⁻¹ are generally consumed right after the start of the experiment (see Figure 4.5a-c).

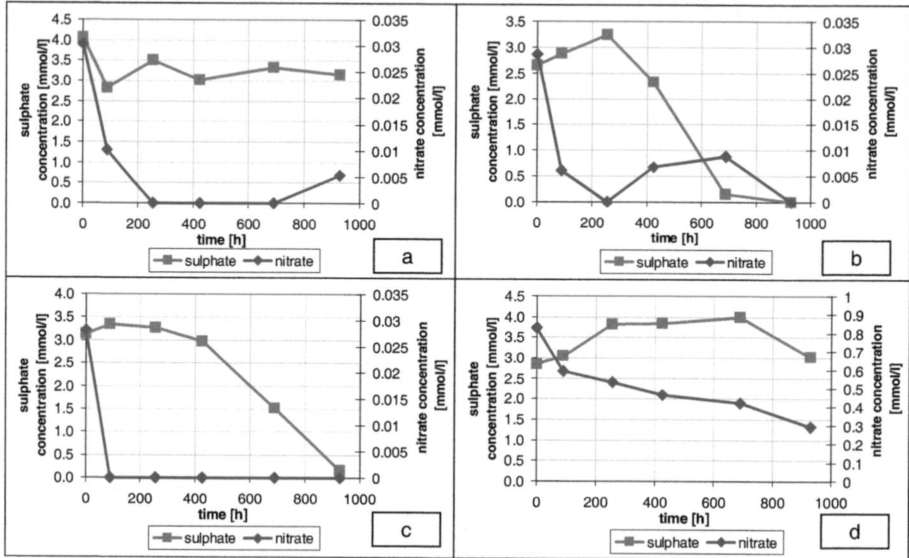

Figure 4.5: Nitrate and sulphate concentrations for the batch experiments: a) reference sample; b) EHC sample; c) glucose sample; d) nitrate sample

If no nitrate and no further carbon addition are added to the batch, indigenous soil microorganisms are able to convert PCE and TCE with a low rate and in a sequential order (see reference sample, Figure 4.4a). After depletion of TCE the concentration of cis-DCE remains constant on a high level. It can be deducted from the anion concentration development (see especially Figure 4.5b+c) that sulphate is consumed in this phase due to sulphate reduction and that cis-DCE is dechlorinated only if sulphate concentrations tend to zero.

When comparing the performance of the two degradation-enhancing substances EHC and glucose it becomes apparent that EHC leads to higher degradation rates than glucose. This refers not only to the degradation of TCE (see Figure 4.4b+c), but also to the consumption of sulphate (Figure 4.5b+c).

The conversion rates $\left(\dfrac{\Delta C_i}{\Delta t}\right)$ of the different experiments are summarised in Figure 4.6. Where possible (i.e. degradation of the respective compound occurred over several sampling times), error bars were added to show the inherent deviation. However, the large intervals between sampling points make this interpretation difficult and thus, the depicted values can only be regarded as first estimates.

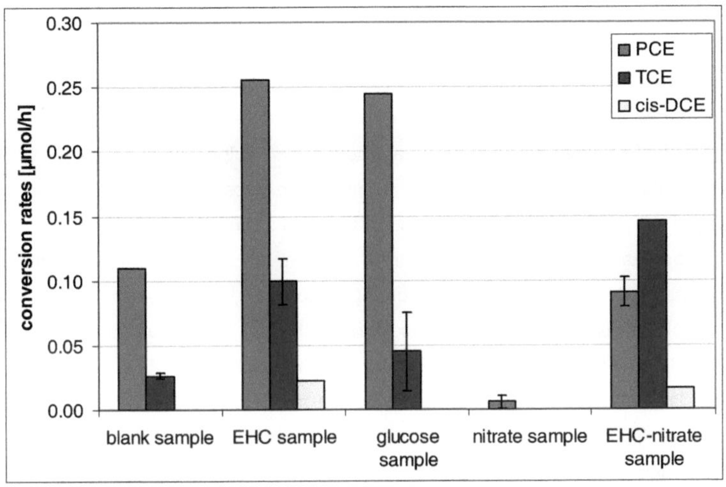

Figure 4.6: Approximated conversion rates of chlorinated ethene degradation in the batch experiment

Further, first order degradation rates could be calculated for some of the experiments. Due to the fact that this method requires three or more data points for an approximative calculation, it could only be performed for TCE in some of the samples. The underlying technique is given by integration of equation (2.9): plotting logarithmised concentration values over time followed by calculation of the slope of the corresponding regression line leads to first-order degradation rates for the compound according to chapter 2.2.5. Table 4.3 summarises these rates for some of the TCE experiments.

Table 4.3: Calculated first-order degradation rates for TCE in selected experiments

Experiment	first-order degradation rate [h^{-1}]
Reference sample	$2.3 \cdot 10^{-3}$
EHC sample	$1.19 \cdot 10^{-2}$
Glucose sample	$1.3 \cdot 10^{-3}$

Summarising the outcomes of the batch test experiment, it can be emphasized that reductive dehalogenation occurs in a strong sequential reaction order which includes the two inorganic electron acceptors nitrate and sulphate. The order deduced in this experiment is given as follows: nitrate reduction → PCE degradation → TCE degradation → sulphate reduction → cis-DCE degradation. Each of these reaction steps seems to inhibit the subsequent reaction to a certain amount. In case of nitrate, the inhibition is assumed to be quite strong due to the observation that no contaminant degradation occurred in the nitrate sample. In the other experiments even 0.03 mmol·l^{-1} of nitrate was consumed in the first hours prior to PCE decomposition.

With respect to transferability towards natural conditions in the aquifer, temperature dependency has to be additionally introduced into the evaluation of reaction kinetics. This phenomenon is considered by van't Hoff's law, which states that an increase in temperature of 10°C approximately doubles to quadruples reaction rates in biological systems. Due to the fact that the experiment described here was conducted at room temperature (~20°C) and the typical aquifer temperature is only 10°C, it might be assumed that the resulting r eaction rates need to be decreased by a factor of two to four in order to fit to natural conditions in the groundwater.

4.2.3 Modelling of the batch test experiment

In a subsequent step the obtained experimental data was utilised to setup a reaction model and to characterise parameters involved in the degradation of chloroethenes. For modelling purposes the simulation software Berkeley Madonna 8.3.18 (http://www.berkeleymadonna.com/) was used. This tool allows the implementation of reaction equations and solves these differential equations providing different methods. Here, the identification of reaction parameters is unaffected by flow and transport processes.

The reaction kinetic of chloroethene degradation chosen for this approach was of first-order type, but moreover, additional terms representing the inhibiting effect of electron acceptors were applied. Due to the fact that only nitrate and sulphate concentrations were determined during the analyses, only these two chemicals were taken into account for extension of the reaction equations. Where available, parameters were taken from measurement results (e.g. Table 4.3) and were used for the setup of the model. The typical reaction equation representing the decay of a chloroethene compound is given by equation (4.1).

$$v = \frac{dC_k}{dt} = -v_{k,max} \cdot C_k \cdot \left[\left(\frac{C_{NO3}}{K_{I,NO3} + C_{NO3}} \right) \cdot \left(\frac{C_{SO4}}{K_{I,SO4} + C_{SO4}} \right) \right] \tag{4.1}$$

$v_{k,max}$: first-order degradation rate of compound k

C_k: concentration of compound k

$K_{I,NO3}$, $K_{I,SO4}$: inhibition constant of nitrate and sulphate, respectively

C_{NO3}, C_{SO4}: concentration of nitrate and sulphate, respectively

Additionally, decay of nitrate and sulphate was approximated by a first-order decay based on measuring data as shown in Figure 4.5. Parameters involved in the degradation of chlorinated ethenes are given in Table 4.4

Table 4.4: Summary of reaction rates and inhibition constants used in the batch test simulations

Compound	first-order reaction rate [h^{-1}]	Nitrate inhibition constant $K_{I,NO3}$ [µmol·l^{-1}]	Sulphate inhibition constant $K_{I,SO4}$ [µmol·l^{-1}]
PCE	$1 \cdot 10^{-2}$ - $8 \cdot 10^{-1}$	0.5	-
TCE	$1.5 \cdot 10^{-3}$ - $3 \cdot 10^{-2}$	0.5 - 1	-
cis-DCE	$1 \cdot 10^{-4}$ - $1 \cdot 10^{-2}$	0.5	1 - 10
VC	$1 \cdot 10^{-3}$ - $1 \cdot 10^{-2}$	0.5	1 - 10

Using the above given parameters the simulation was performed and the concentration results were plotted versus time. The outcomes for reference sample, EHC and glucose experiment as well as the nitrate experiment are shown in Figure 4.7.

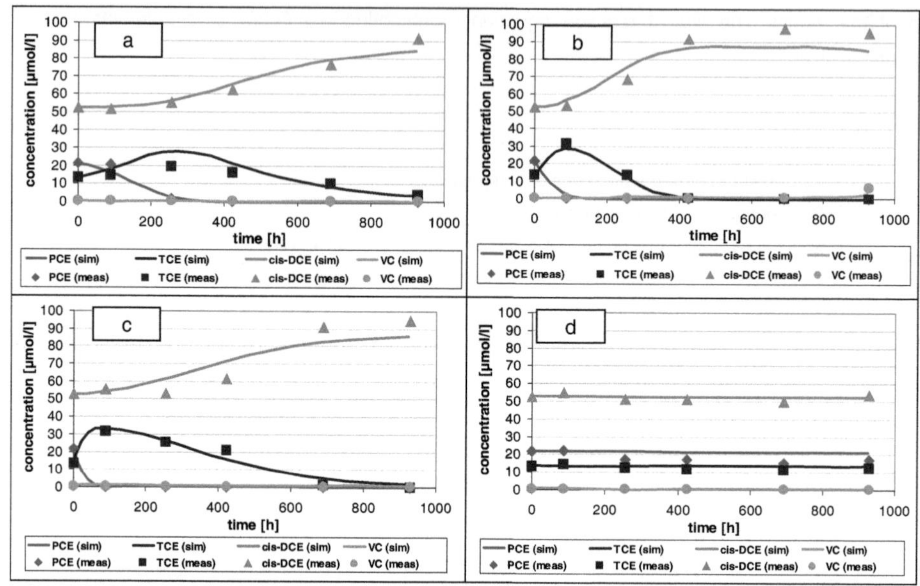

Figure 4.7: Simulation results (sim) and measurement data (meas) of the batch test: a) reference sample; b) EHC sample; c) glucose sample; d) nitrate sample

It becomes obvious from Figure 4.7 that modelling contaminant decay using inhibition terms for different electron acceptors delivers reasonable results for all performed simulations. Simulated and measured concentration values match very well with only few exceptions. This outcome is thus seen as a proof of concept and is henceforward applied for the more comprehensive groundwater models described in chapter 5.

In detail, the degradation rates of TCE derived from the experiment and the simulation (compare Table 4.3 and Table 4.4) are in good accordance with each other. The range of TCE reaction rates is similar in both cases. Additionally, the choice of nitrate and sulphate inhibition parameters seems to fit quite well, leading to application of these values in the models described in chapter 5.3.2. Conversion and application of van't Hoff's law results in data ranging from $0.22 - 43.8 \ y^{-1}$ for the first-order degradation rates derived from the experiments. When comparing these values with literature data (see Figure 2.5), which range around $1 \ y^{-1}$ (compare Figure 2.5), it gets clear that both are in the same order of magnitude. This observation gives evidence for the suitability of the experimental setup, which appears reasonable to examine chloroethene degradation in a laboratory setting and to derive degradation and inhibition constants.

4.3 Characterisation of chloroethene-degrading bacteria

4.3.1 Materials and methods

As discussed in chapter 2.1.2 different soil bacteria are capable of degrading chloroethenes. However, most of these autochthonous, ubiquitous bacteria stop the degradation chain at cis-DCE and do not continue contaminant decomposition under reducing conditions (e.g. *Dehalobacter restrictus* and *Desulfuromonas michiganensis*) (Watts et al., 2005). Up to now only the strain *Dehalococcoides ethenogenes 195* is known to be capable to completely degrade chloroethenes to ethene (Hendrickson et al., 2002). Hence, the experiment described here attempted to detect this specific microorganism in groundwater samples. The presence of *D. ethenogenes* might provide an indication of the natural degradation potential in the soil. Another aim of this experiment was to prove the correlation of presence of contaminant degrading bacteria with certain redox potential, i.e. whether certain environmental conditions influence the bacterial community with regard to dehalogenating bacteria. From the theoretical point of view it is expected that dehalogenating bacteria prefer reducing conditions and are thus predominantly present in the downstream part of the experimental area.

Metagenomic techniques were employed to identify the presence of potential pollutant-degrading bacteria. Via the polymerase chain reaction (PCR; Saiki et al., 1985) using organism-specific primers, amplification of DNA was performed and the amplified DNA was afterwards stained and detected in an agarose gel. The detection of *D. ethenogenes* via molecular genetic methods was performed in groundwater samples taken from the case study area described in chapter 3. Seven samples from observation wells along the contaminant plume were collected (see Figure 4.8), each comprising different environmental conditions: in the upstream part of the domain where the pollutant source was supposedly located, nitrate-reducing conditions dominate, but contaminant concentrations have already declined (well B5). Further downstream a shift towards iron(III)-reducing conditions occurs and PCE as well as TCE concentrations increase (wells B16 and B12). In the subsequent wells high cis-DCE concentrations were analysed and thus it is assumed that PCE and TCE degrading bacteria might be found here (SB1 and B14). The wells in the downstream part (TUBS3 and TUBS1) exhibit the lowest redox potential and also decreasing concentrations of cis-DCE and traces of VC.

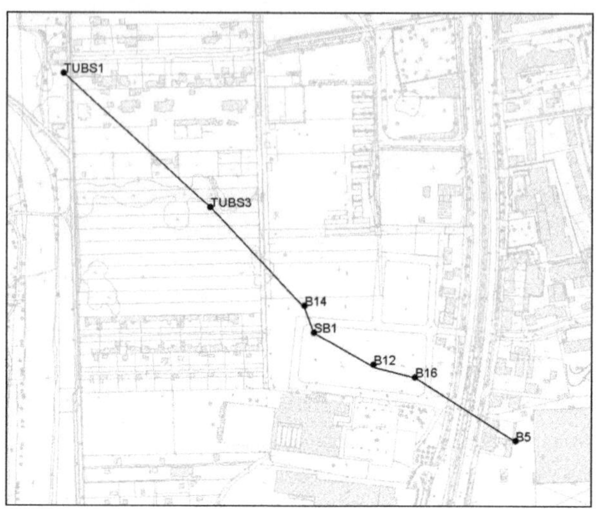

Figure 4.8: Observation wells along the transect line in the Schützenplatz area sampled for detection of dechlorinating microorganisms

DNA was extracted from the groundwater samples according to the protocol described in the appendix (chapter 11.2). Samples with low turbidity (i.e. from wells B5, B16 and TUBS3) were centrifuged (a) and additionally filtered (b) at the start of the DNA purification in order to gain a higher amount of solid matter for the next purification steps. Samples from the wells B12, SB1, B14 and TUBS1 were centrifuged only and further processed according to the appended protocol for DNA extraction and purification.

The extracted DNA was tested for several degradation-specific enzymes of different microbial strains (reductive dehalogenases) as well as for specific conserved rRNA-sequences typical for dehalogenating bacteria. Primers for the following sequences were designed (see Table 11.5) and PCR was performed (operational procedure described in the appendix, chapter 11.2):

- tceA: dehalogenase isolated from *D. ethenogenes 195* (Magnuson et al., 2000): expected fragment length: 260 bp; main enzyme substrates are TCE and cis-DCE
- bvcA: dehalogenase isolated from *Dehalococcoides* sp. strain BAV1 (Krajmalnik-Brown et al., 2004): expected fragment length: 295 bp; main enzyme substrates are all DCE isomers and VC
- vcrA: dehalogenase isolated from *Dehalococcoides* sp. strain VS (Muller et al., 2004): expected fragment length: 300 bp; main enzyme substrates are all DCE isomers and VC

- combined approach for *Dehalobacter*-specific dehalogenases as described e.g. in Villemur et al. (2002): expected fragment length: 230 bp; main enzyme substrates are PCE and TCE
- 728fmod/792r: conserved 16S-rRNA region of dehalogenating species: expected fragment length: 126 bp
- 728fmod/1172rmod: conserved 16S-rRNA region of dehalogenating species: expected fragment length: 435 bp
- 774f/1172rmod: conserved 16S-rRNA region of dehalogenating species: expected fragment length: 328 bp

All rRNA-primers given here were aligned according to the Ribosomal Database Project (http://rdp.cme.msu.edu/) using the software MUSCLE (Multiple Sequence Alignment) of the software package SeaView (Gouy et al., 2010).

4.3.2   Results of the experiment

The DNA-purification revealed a satisfying concentration in almost all of the environmental samples. DNA-concentrations ranged from 6 to 72 ng·dl^{-1}. DNA-samples of *Dehalococcoides ethenogenes* 195 (provided by Ivonne Nijenhuis, UFZ Leipzig) showed a positive PCR-result for tceA-gene and all 16S-rRNA-samples, indicating that the PCR-procedure is operational. As expected for the other PCR-products, no DNA bands were detected with the *D. ethenogenes* DNA template. However, in the environmental samples no PCR-product was found using the above described primer-sets except for the 774f/1172rmod 16S-rRNA-primers. The corresponding gel-photo (Figure 4.9) indicates DNA bands at the expected fragment length of around 330 bp. Figure 4.10 shows a comparison of redox potential and cumulated chloroethene concentrations (CHC) at the respective wells along the transect.

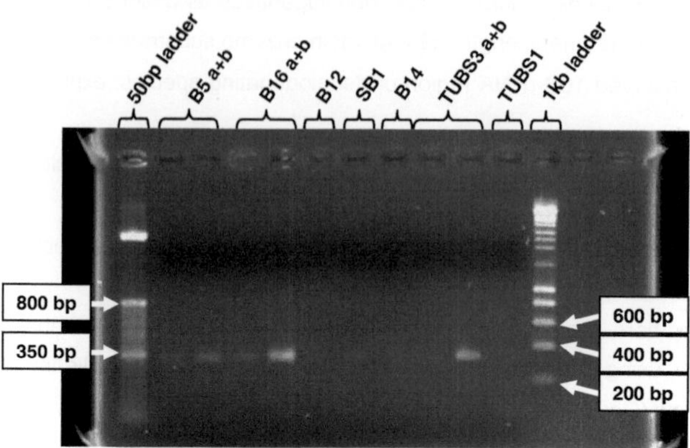

Figure 4.9: Gel-photo of the PCR with 774f/1172rmod-primers

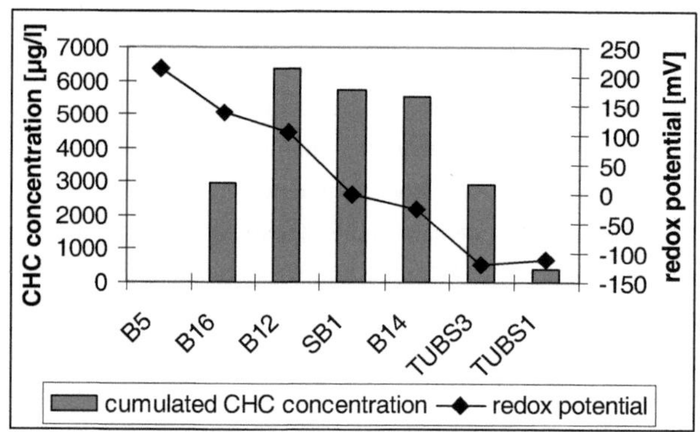

Figure 4.10: Measurement data for chloroethene concentrations and redox potential at the observation wells sampled for the characterisation of dehalogenating bacteria

Sequencing of the respective DNA revealed homology with the PCB-dechlorinating bacteria DF-1 and o-17 described in Watts et al. (2005). These bacteria are also capable of dehalogenating PCE and TCE to cis-DCE, but no further degradation to ethene is catalysed. This fact is in agreement with the observed accumulation of cis-DCE in the case study area. Sequence similarity to *D. ethenogenes 195* is only 89%, indicating that these species are distinct on genus level or higher (Miller et al., 2005). However, along with the result of the *D. ethenogenes 195* positive control it could be shown that the selected 16S-rRNA primers are capable to align with a wide variety of dehalogenating bacteria-DNA and thus, detection of a wide range of these species was possible. Unfortunately the detection of DNA bands from the environmental samples shows no correlation with redox potential or contaminant distribution. As it becomes obvious from Figure 4.9 the dechlorinating bacteria were detected in the upstream part (B5, B16) where contaminant concentrations already declined and redox potential is around ~+200 mV, as well as in the downstream part (TUBS3) where cis-DCE concentration are high and redox conditions show a potential of around ~-100 mV (compare Figure 4.10). The initial assumption that dehalogenating bacteria prefer reducing conditions could not be confirmed by the experimental results. Besides that it could not be proven that high contaminant concentrations correspond with occurrence of dechlorinating bacteria.

Given the assumption that DNA-purification and PCR-procedure worked faultless with the environmental samples, no *Dehalococcoides* species were found in any of the samples. It seems reasonable to assume that biodegradation of chloroethenes is catalysed by the detected DF-1- and o-17-strain as well as other species, which were not detectable with the primers used in this experiment. However, concluding from the experiment conducted here a bioaugmentation of heavily contaminated areas with an enrichment culture consisting of *D. ethenogenes* might be a feasible method to induce pollutant degradation beyond cis-DCE.

5 Finite Element groundwater modelling

The following chapter deals with the construction and calibration of a finite element groundwater model demonstrated for a field scale study area (chapter 5.1), the conduction of a sensitivity analysis for important flow, transport and reaction parameters and finally the introduction of a concept to assess uncertainties in modelling of contaminated aquifers. This approach is based on a MC simulation, which is employed for the derivation of contaminant-specific spatial and temporal probabilities of occurrence (chapter 5.2). The final aim of this method, however, is to build up a basis for a probabilistic risk assessment approach including uncertainty evaluation. The underlying conceptual advance is shown in Figure 5.1. While the first step (test field exploration) is further explained in chapter 3, the entire model-related topics (except for the risk assessment approach) are described in this chapter. For groundwater modelling purposes the Finite Element software Feflow 5.412 (DHI-WASY; Diersch, 2009) is used.

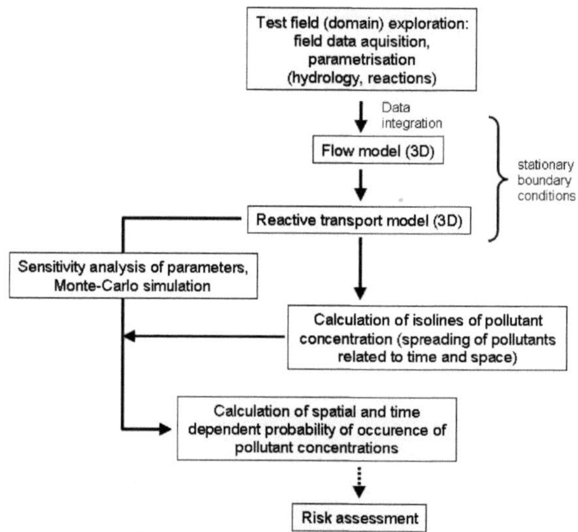

Figure 5.1: Conceptual approach for calculation of contaminant-specific probabilities of occurrence for risk assessment purposes

Further, a parameter optimisation using the MC technique is established and applied for the mentioned case-study (chapter 5.2). The outcomes of this approach suggested a modification of the formerly used first-order degradation kinetic. Based on the results of the batch experiment described in the previous section this implementation was performed for different model setups and validated with field data (chapter 5.3).

5.1 Steady-state Finite Element model

5.1.1 Definition of model domain and initial conditions

Several sequential working steps are carried out to develop a groundwater model for reactive transport simulations. The first step in model construction is the implementation and calibration of the hydraulic model. Hence, groundwater head measurements are conducted and utilised for the interpolation of a head isoline map. According to this isoline map the model domain is defined in such a way that model borders run along the head isolines or cut them perpendicularly. The intention behind utilisation of this procedure is a minimum mismatch of the model's water balance.

In a second step borehole drilling profiles are evaluated for the occurrence of different soil layers (see Table 3.1). The thickness of the different layers is transferred into the model and interpolated over the entire domain. The stratigraphic data is exemplarily shown for a cross-section in Figure 5.2 (with reference to Figure 3.1).

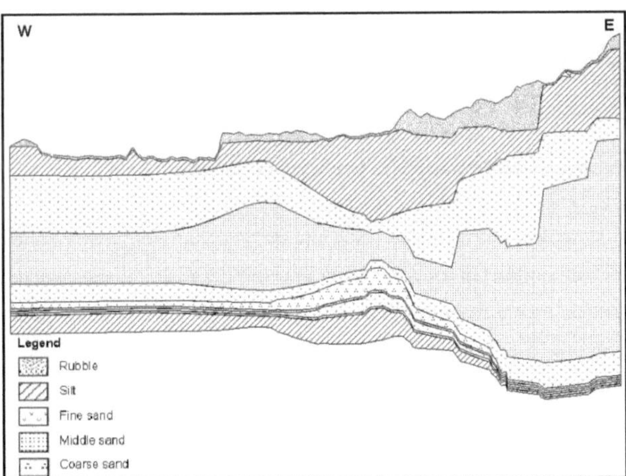

Figure 5.2: Exemplary illustration of the layer model implemented by means of petrographic data (vertically exaggerated); cross-section shown here according to horizontal line in Figure 3.1

Based on the spatial extent of the model domain and the stratigraphic data, the FE mesh was generated by means of triangular elements. The mean horizontal element longitude is around 6.5 m, leading to Peclét numbers between 0.3 and 1.2, which is adequate in terms of numerical stability. Due to high concentration gradients an additional mesh refinement is necessary for the contaminant source area. This refinement was carried out until numerical stability is considered to be sufficient. The entire model finally consists of almost 200,000 triangular elements and about 10,000 nodes per slice, covering a total volume of $4.6 \cdot 10^6$ m³.

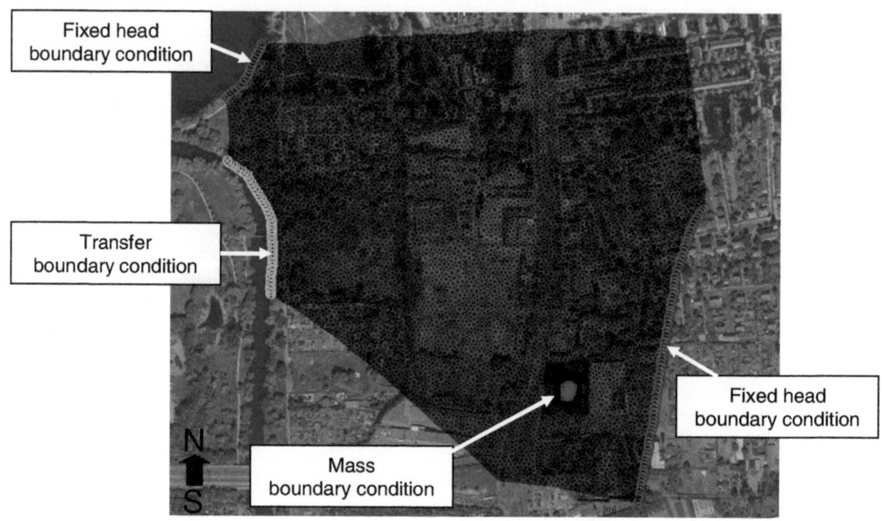

Figure 5.3: Model domain with FE mesh and boundary conditions

Additionally, the steady-state flow model builds upon several initial and boundary conditions. At the western border the mean water level of the river Oker and the Ölper Lake is implemented as fixed head derived from measured mean groundwater levels (Dirichlet BC), while at the eastern border a fixed head is interpolated. The groundwater head is assessed by several water table measurements of 26 observation wells in the area.

Unfortunately, the exact location of the contaminant source zone as well as the pollutant infiltration period is unknown and could not entirely be retrieved from inspection of records. Thus, reverse modelling of the current pollutant distribution was performed in order to approximate the most probable location of the source area. Further, it is assumed that the pollution of the aquifer started in 1960 and lasted for 15 years. Over the years PCE has been infiltrated into the subsurface with a maximum solubility of 160 mg·l⁻¹ (time-dependant Dirichlet mass BC).

In 1974 German authorities restricted usage and prohibited disposal of chlorinated ethenes by law (Bundes-Immissionsschutzgesetz (BImSchG, 1974) and appending Bundes-Immissions-schutzverordnungen (BImSchV)). Therefore, this date was chosen for formulation of the source release boundary conditions. In particular, the infiltration of PCE in the model stops in 1975 and no further infiltration occurs from this time onwards. Thenceforward, the pollutant plume is only subject to transport and reaction processes. Modelling the current contaminant plume extent (present date) thus refers to 50 years of simulation time.

5.1.2 Setup and calibration of the hydraulic model

As discussed in the previous chapter a constant head based on groundwater head measurements is assigned for the stationary model. A head of 67.09 m is applied at the eastern border of the model domain and 66.3 m at the lake Ölper. This corresponds to a hydraulic gradient of approximately 1.2 ‰. A transfer rate of $99.36 \cdot 10^{-4}$ d^{-1} is set at the transfer boundary condition along the river Oker, referring to an almost impermeable colmation layer ($K_f = 10^{-8} - 10^{-9}$ $m \cdot s^{-1}$) with a thickness of 0.5 m. Additionally a surface infiltration of $1.7 \cdot 10^{-4}$ $m \cdot d^{-1}$ is used for unsealed areas to specify precipitation.

After flow parameters such as hydraulic conductivity are determined according to soil properties, the model was run to check the water balance. The results (Figure 5.4) reveal that the main water fluxes flow through the head boundary conditions and precipitation only plays a minor role. The water imbalance of the model is close to zero, representing an acceptable convergence towards natural conditions.

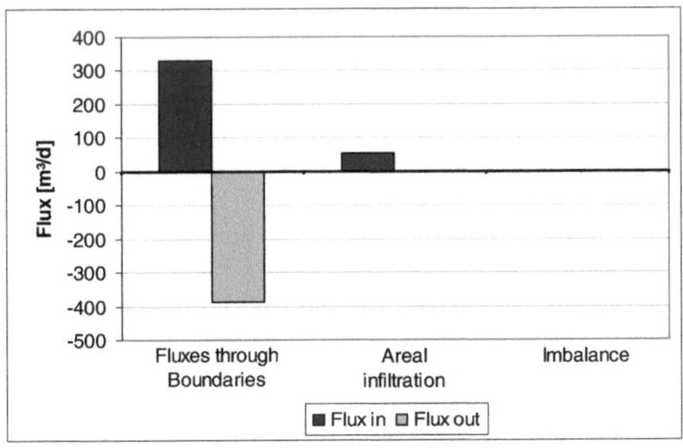

Figure 5.4: Water balance of the steady-state groundwater model (positive values: inflow; negative values: outflow)

A comparison of measured and simulated groundwater head at the observation wells is depicted in Figure 5.5. In a preliminary model approach the measurement data was underestimated by the flow model (see appendix: Figure 11.3). Although this underestimation was only in the range of few centimetres, additional modifications of the flow boundary conditions (increase of groundwater head BC by 8 cm) were integrated in the model. Results are in better agreement of both, simulated and measured data (Figure 5.5).

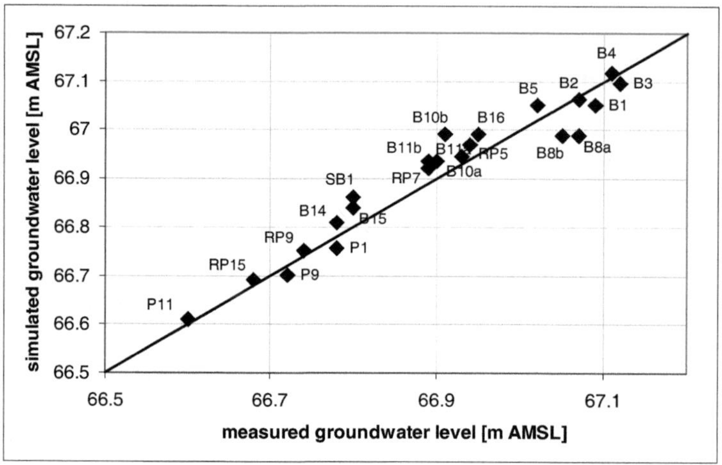

Figure 5.5: Scatter plot of measured versus simulated groundwater head (referring to meters above mean sea level (m AMSL))

Based on the calibrated flow model the implementation of the transport and reaction model was performed.

5.1.3 Setup and calibration of the transport and reaction model

Using the steady-state flow model described in the previous chapter a transport model is implemented using transport parameters determined in chapter 4.1. In addition, first-order degradation kinetic is applied. The kinetic is based on literature values and the experiments described in chapter 4.2. The parameter set for this preliminary reference model is given in the appendix. Stoichiometric coefficients for each reaction can also be found there (Table 11.7).

Similar to the calibration of the flow model, a validation is performed with the concentration data determined during measurement campaigns. The scatter plot showing the comparison between both data sets is depicted in Figure 5.6. Caused by the large range of concentration values, a double-logarithmic depiction had to be chosen for this figure. Additional model validation measures, e.g. break-through curves for single observation wells, were not realisable due to the scarce temporally-resolved measurement data. Furthermore the complex stratigraphic soil composition in the experimental domain limits the informative value of such procedures.

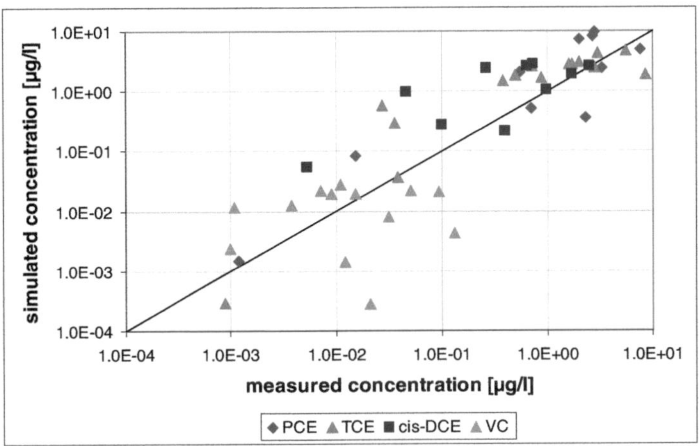

Figure 5.6: Scatter plot of measured versus simulated concentrations

The extension of the flow model by transport and reaction processes generally yielded acceptable simulation results (Figure 5.6). However, some deviations of concentration data from the bisecting line are obvious and may be explained by groundwater head oscillations and other reasons as described below.

While the simulation software Feflow 5.412 includes the parameter estimation software PEST for estimation and adaptation of parameters, this tool is only dealing with flow processes. In contrast to that, the manual adjustment of transport and reaction parameters is regarded to be highly complex but inevitable, as no other sophisticated solutions are available to date. Furthermore, flow and transport parameters are strongly coupled between each other, e.g. an increase in PCE degradation rate does not only influence the PCE concentrations, but affects concentrations of all subsequent compounds. With respect to these issues, an objective way to compare and grade the different reactive transport models was necessary to develop. This was performed by means of a MC approach as described in the following section.

In addition, fluctuation in measurement data, e.g. caused by sampling inaccuracies or due to groundwater head oscillations, introduces a further element of uncertainty. Finally, due to the high number of parameters influencing the model outcomes and the heterogeneities of the natural system, which cannot be represented by modelling tools, an ideal match of both, measured and simulated concentrations is nearly impossible. In total the calculated deviations can be deemed satisfactory for the subsequent development of uncertainty assessment tools and thus, the efforts of establishing transport and reaction processes in the model are presumed successful. A potential approach to solve the described difficulties is presented in the following chapters. Here a mathematically improvement of transport and reaction parameters based on the MC technique is performed for the test area (chapter 5.2.3). Additionally an enhanced reaction kinetic is developed in order to further improve the model outcomes (chapter 5.3).

5.2 Model results and model optimisation

Based on the steady-state groundwater model described in the previous chapter, different approaches for uncertainty assessment of the reactive transport processes as well as a parameter optimisation procedure was conducted. The uncertainty evaluation which is performed by means of a sensitivity analysis for several flow, transport and reaction parameter is described in chapter 5.2.1. For the uncertainty assessment a MC simulation was used which is further discussed in chapter 5.2.2. Results of both sections are also published and further discussed in Greis et al. (2011).

In addition a parameter optimisation approach using MC methods for the field-site area discussed in chapter 3 is set up and tested in chapter 5.2.3. The uncertainty calculation was necessary due to the scarce data base regarding concentration measurement data and aims to a better prediction of contaminant plume propagation.

5.2.1 Sensitivity analysis

Based on the model established in the previous chapter, a sensitivity analysis is performed for seven important flow and transport parameters in order to evaluate their influence on the resulting model outcomes. The tested parameters are porosity ε, hydraulic conductivity K_f (see chapter 2.2.1), longitudinal and transversal dispersivity D_L and D_T (see chapter 2.2.2), diffusion coefficient D_D (see chapter 2.2.3), reaction rate v (see chapter 2.2.5) and linear adsorption isotherm coefficient K_D (, see chapter 2.2.4). The available literature data for each parameter is analysed and on this basis preliminary extreme value scenarios for evaluation of the upper and lower bounds of each parameter are performed. Results of the sensitivity analysis are exemplarily depicted for diffusion coefficient and hydraulic conductivity in Figure 5.7.

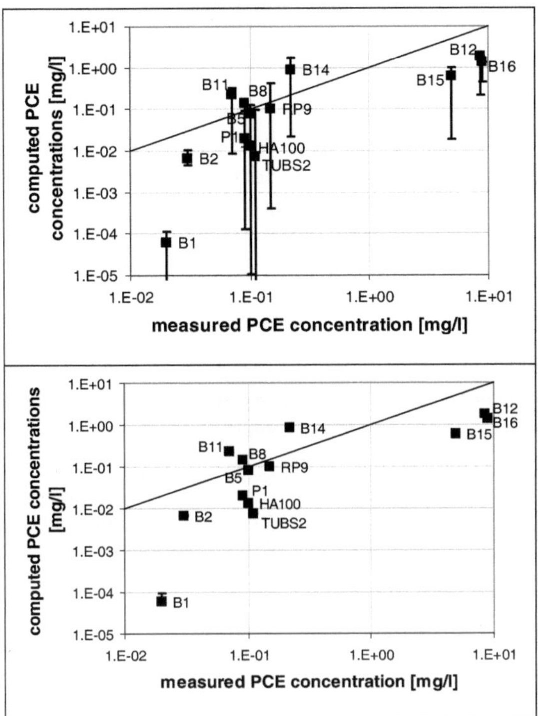

Figure 5.7: Comparison between variation of diffusion coefficient D_d (left) and hydraulic conductivity K_F (right) on transport of PCE (Greis et al., 2011)

It could be shown in the sensitivity analysis, that diffusion and porosity have minor influence on contaminant transport, while dispersivity, adsorption, reaction rate and hydraulic conductivity have major impact on transport behaviour for the example study site. Despite the large interval of variation of the diffusion coefficient over three orders of magnitude ($1 \cdot 10^{-9} - 1 \cdot 10^{-6}$ m$^2 \cdot$s^{-1}), differences in resulting contaminant concentrations are negligible. On the other hand even small changes in hydraulic conductivity (±50% of the value implemented in the base model) lead to extremely differing results (see Figure 5.7).

5.2.2   Monte-Carlo simulation for calculation of probability isolines

As a result of the preceding sensitivity analysis the range of parameters for a subsequent MC simulation has been defined. This analysis is carried out to identify the range of stochastical parameter variation for the MC simulation. It is assumed that all parameters used for the MC simulation are Gaussian distributed and that the parameter values for the MC simulation are in the range of plus/minus two times the standard deviation ($\pm 2\sigma$) of the results identified by the sensitivity analysis (i.e. 95.45% of the generated values are in the interval). Stochastically independent, Gaussian-distributed random variables are generated using the Mersenne twister algorithm (as implemented in Mathworks Matlab 2009b).

Table 5.1: soil parameters used in the base model and standard deviation σ of parameters applied for the MC simulation

layer	soil type	hydraulic conductivity [10^{-4} m·s^{-1}]	porosity [-]	longitudinal dispersivity [m]	transversal dispersivity [m]
1	rubble	5 ± 1.25	0.25 ± 0.063	13.2 ± 3.3	1.98 ± 0.5
2	silt	1.16 ± 0.29	0.05 ± 0.013	9 ± 2.25	1.35 ± 0.34
3	fine sand	4.63 ± 1.16	0.1 ± 0.025	12 ± 3	1.8 ± 0.45
4	medium sand	23.15 ± 5.79	0.12 ± 0.03	11.4 ± 2.85	1.71 ± 0.43
5	fine sand	4.63 ± 1.16	0.1 ± 0.025	12 ± 3	1.8 ± 0.45
6	coarse sand	30 ± 7.5	0.17 ± 0.043	15 ± 3.75	2.25 ± 0.56
7	medium sand	23.15 ± 5.79	0.12 ± 0.03	11.4 ± 2.85	1.71 ± 0.43
8	fine sand	4.63 ± 1.16	0.1 ± 0.025	12 ± 3	1.8 ± 0.45
9	coarse sand	30 ± 7.5	0.17 ± 0.043	15 ± 3.75	2.25 ± 0.56
10	silt	1.16 ± 0.29	0.05 ± 0.013	9 ± 2.25	1.35 ± 0.34

Table 5.2: contaminant parameters used in the base model and standard deviation σ of the parameters generated for the MC simulation

compound	K_D-value [10^{-3}]	reaction rate [10^{-9} s^{-1}]	diffusion coefficient [10^{-9} m²·s^{-1}]
PCE	4.44 ± 1.11	10 ± 3.75	50 ± 25
TCE	4.44 ± 1.11	10 ± 3.75	50 ± 25
cis-DCE	3.89 ± 0.97	1 ± 0.188	50 ± 25
VC	3.33 ± 0.83	10 ± 3.75	50 ± 25

Using the parameters listed above, the MC variants of the transport model are parameterised and contaminant propagation is calculated. The concentration outcome for each of the contaminant compounds (PCE, TCE, cis-DCE and VC) at every node of the FE grid is exported for all the MC variants. Certain threshold values (0.1, 0.5 and 1 mg·l^{-1} for PCE, TCE and cis-DCE; 0.02, 0.05 and 0.1 mg·l^{-1} for VC) are chosen arbitrarily and the amount of nodes exceeding these thresholds is counted. Summation of variants exceeding threshold concentrations at a specific node and comparison with the overall number of simulations deliver a percentage of contaminant-occurrence, which is referred to as probability of occurrence of specific concentration of this compound at the respective node (e.g. >50%, >80% or >90%). Via graphical analysis, probability isolines for each concentration of a compound are constructed.

A correlation analysis is performed in order to evaluate an appropriate number of MC simulations, which is sufficient for a substantive prediction of pollutant transport behaviour. The results are illustrated in Figure 5.8, showing the correlated concentration results of each model node for different numbers of MC variants: for instance V50-60 means that results of the first 50 variants are compared to those of the first 60 variants. The outcome reveals an increasing correlation with increasing number of variants until 80 variants. Beyond this number, the correlation coefficient stagnates, concluding that 100 MC simulations are sufficient for calculating contaminant-specific probabilities of occurrence.

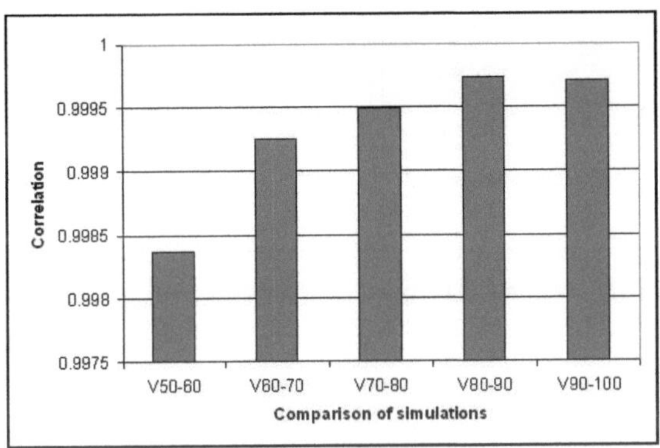

Figure 5.8: Correlation analysis to derive a sufficient number of MC simulations (Greis et al., 2011)

Further results of the MC simulation are evaluated by graphical analysis. The obtained probability isolines for different pollutant concentrations are represented in Figure 5.9.

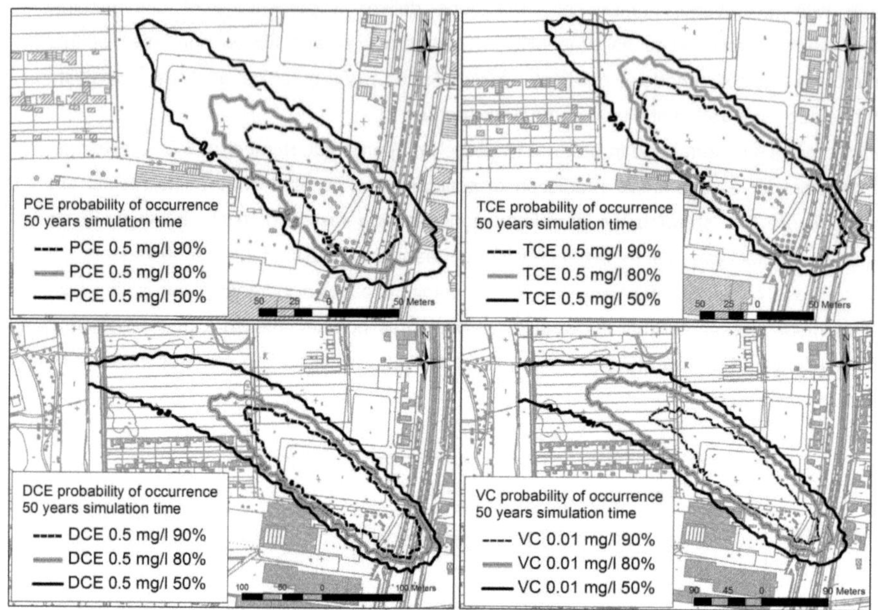

Figure 5.9: Isoline visualisation of the compound specific probabilities of occurrence, referring to 50 years simulation time (Greis et al., 2011)

The graph (Figure 5.9) exhibits the results of 50 years simulation time. The isolines stand for 50%, 80% and 90% probability that 0.5 $mg \cdot l^{-1}$ concentration of PCE, TCE, cis-DCE and 0.02 $mg \cdot l^{-1}$ of VC will be exceeded.

It becomes obvious, that the tailing of lower substituted chloroethenes (cis-DCE and VC) tends to be more distinctive than that of higher substituted ethenes (PCE, TCE). In particular, the different isolines of VC have a larger distance from each other in comparison to the TCE or PCE isolines. This effect can be interpreted by a higher dependency from variation in reaction rates: the concentration of PCE, for example, is only dependent on its own degradation rate (besides transport and flow parameters), while VC concentrations are also dependant on the varying degradation rates of the previous compounds. In consequence, parameter dependency and thus uncertainty is higher for substances at the end of the reaction chain.

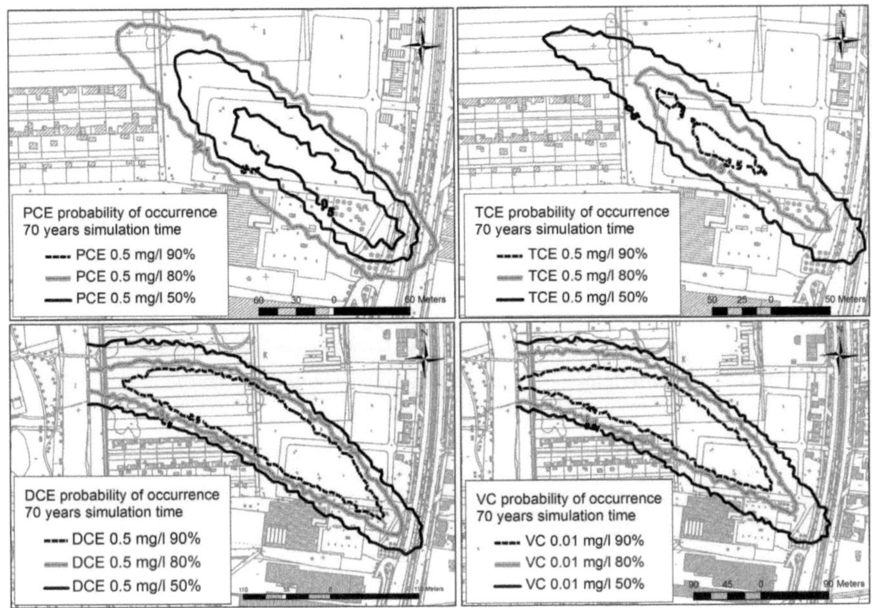

Figure 5.10: Isoline visualisation of the compound specific probabilities of occurrence, referring to 70 years simulation time (Greis et al., 2011)

Additional results for a 70 years simulation time are shown in Figure 5.10. Comparison of both simulation periods shows, that contaminant spectrum changes more towards lower substituted chlorinated ethenes. Respective areas of cis-DCE and VC pollutants are getting more distinctive in the simulation. On the other hand, PCE and TCE affected areas are declining. However, increasing concentrations of pollutants are expected to be spilled into the surface water river Oker or lake Ölper.

5.2.3   Model optimisation using Monte-Carlo method

In the previous chapter 5.1, simulation outcomes were assessed by comparison of simulated concentrations at 19 observation wells with the corresponding measurement data using a scatter plot depiction (e.g. see Figure 5.6). In order to evaluate the simulation outcomes objectively and in a quantitative way, the quadratic deviation of simulated and measured data for all 19 observation wells is utilised to assess the respective model quality. The resulting quadratic deviation (entitled here: absolute error) for a single compound is an abstract value which serves for comparison of different model variants. Small deviation values represent a better agreement of simulation and measurement data and are thus tried to be achieved with the model approach. For a single chloroethene the quadratic deviation is calculated according to the following equation (5.1):

$$Err_{abs,k} = \sum_{wells} \left(C_{meas,k} - C_{sim,k} \right)^2$$

(5.1)

$Err_{abs,k}$: absolute error for compound k

C_{meas}: measured concentration of CHC compound

C_{sim}: simulated concentration of CHC compound

If the deviation of the data pairs is summed up for all four chemical compounds (PCE, TCE, cis-DCE, VC), the calculated error is termed cumulated absolute or total error and is described by equation (5.2):

$$Err_{abs} = \sum_{species} \sum_{wells} \left(C_{meas,k} - C_{sim,k} \right)^2$$

(5.2)

Err_{abs}: cumulated absolute error

The concentration of single compounds in the case study ranges over 2-3 orders of magnitude and the maximum concentrations vary from $\sim10^{-1}$ mg·l⁻¹ for VC up to $\sim10^{1}$ mg·l⁻¹ for PCE and cis-DCE. The calculation and comparison of the total error $Err_{abs,k}$ between all contaminant species reveals that compounds with high concentration values contribute to a larger extent to the total error than compounds with low concentrations. To solve this problem the error of each compound is normalised to the error of the respective chemical compound in the original reference model (described in chapter 5.1). Thus, all four substances are subject to optimisation with the same weighting. The resulting error is termed cumulated normalised error:

$$Err_{norm} = \sum_{species} \frac{\displaystyle\sum_{wells} \left(C_{meas,k} - C_{sim,k} \right)^2}{\displaystyle\sum_{wells} \left(C_{meas,k} - C_{org,k} \right)^2}$$

(5.3)

Err_{norm}: cumulated normalised error

C_{org}: simulated concentration of CHC compound in the original model

By minimisation of the Err_{norm} using a MC approach an improvement of the model towards the measured concentration data is supposed to be achieved. Hence, the equation for the normalised error (equation (5.3)) serves as objective function for the following parameter optimisation using MC simulation. The absolute error formula (equation (5.2)) is utilised to evaluate the optimisation approach.

In the simulations, the range of important flow, transport and reaction parameters (as determined by sensitivity analysis, see section 5.2.1) is generated stochastically according to the procedure described in chapter 5.2.2 and the corresponding pollutant concentrations are computed. Similar to the procedure for calculation of probability isolines, the parameters chosen for variation are hydraulic conductivity (K_F), porosity (ε), longitudinal and transversal dispersivity (D_T and D_L), diffusion coefficient (D_D), sorption coefficient (K_D) and reaction rate (v). In total four MC simulations are performed, each based on around 1200 model variants. The parameter variation is assumed to be normal distributed and the standard deviation is chosen according to the parameter values of the original reference model. The standard deviation is decreased with rising number of MC simulation to reflect the overall model improvement (starting with $\sigma_1 = 0.5 \cdot \mu$ $\rightarrow$ $\sigma_2 = 0.25 \cdot \mu$ $\rightarrow$ $\sigma_3 = 0.15 \cdot \mu$ $\rightarrow$ $\sigma_4 = 0.1 \cdot \mu$).

Related to the above given equations (equation (5.2) and (5.3)), the normalised and absolute error of the MC simulation is calculated for all ~1200 model runs of each simulation. The model with the lowest normalised error is chosen as reference model for the subsequent MC run. The original reference model used for the calculation of probability isolines (chapter 5.2.2) also serves as starting point for the model optimisation runs.

Due to the error-minimisation strategy, both kinds of calculated errors are supposed to show a decreasing trend in order to prove the feasibility of this method. The error evolution is shown in Figure 5.11 for the cumulated error values and confirms the aforementioned assumption.

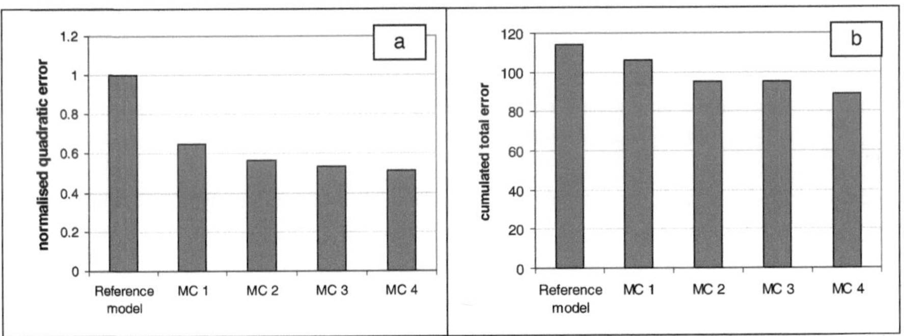

Figure 5.11: Development of cumulated normalised error Err$_{norm}$ (a) and cumulated absolute error Err$_{abs}$ (b) in the MC simulations

A decline of the normalised error to almost 50% is achieved throughout the MC simulations (Figure 5.11a). The error asymptotically approaches a minimum, proving the feasibility of the optimisation approach. This observed asymptotic behaviour was expected for the optimisation problem. However, the cumulated absolute error only declines by almost 22% (from 114 to 89, Figure 5.11b). Nevertheless, a qualitative improvement of the model is still recognisable.

As the cumulated total error Err_{abs} results from summation of the errors of each compound ($Err_{abs,k}$, equation (5.1)), the single compound error data is also depicted in the following graph (Figure 5.12).

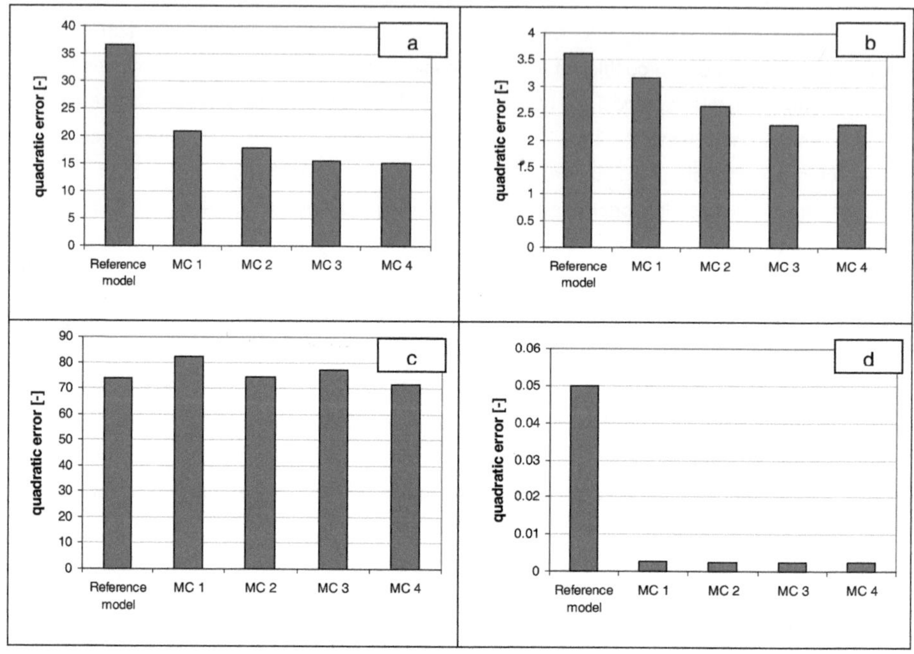

Figure 5.12: Comparison of absolute error development of the four simulated species (a: PCE, b: TCE, c: cis-DCE, d: VC)

As derived from Figure 5.12a+b, error evolution of PCE and TCE follows an asymptotic shape as in the graph for the normalised error (Figure 5.10). Parameter optimisation for these two compounds appears to work well with the measurement data from august 16th and 17th, 2010. For cis-DCE and VC the error development shows a different behaviour. While the error for cis-DCE remains on a continuously high level (Figure 5.11c), VC error rapidly declines at first and remains approximately constant finally (Figure 5.11d). Obviously the optimisation did not work successfully for these substances. For VC one problem may be the result of a rather diffuse, heterogeneous spatial distribution at low concentrations which was determined during the measurement campaigns (see Figure 3.5). For cis-DCE the main difficulties arise from large concentration variations within small spatial distances, which can not be described by conventional models. Additionally, cis-DCE concentrations were the highest among all pollutant species and hence, the calculated absolute error of cis-DCE has major influence on the cumulated error. A sophisticated method to take these circumstances into account was considered after the evaluation of the optimisation approach and resulted in development of a redox-dependant degradation kinetic for contaminant decomposition (described in chapter 5.3).

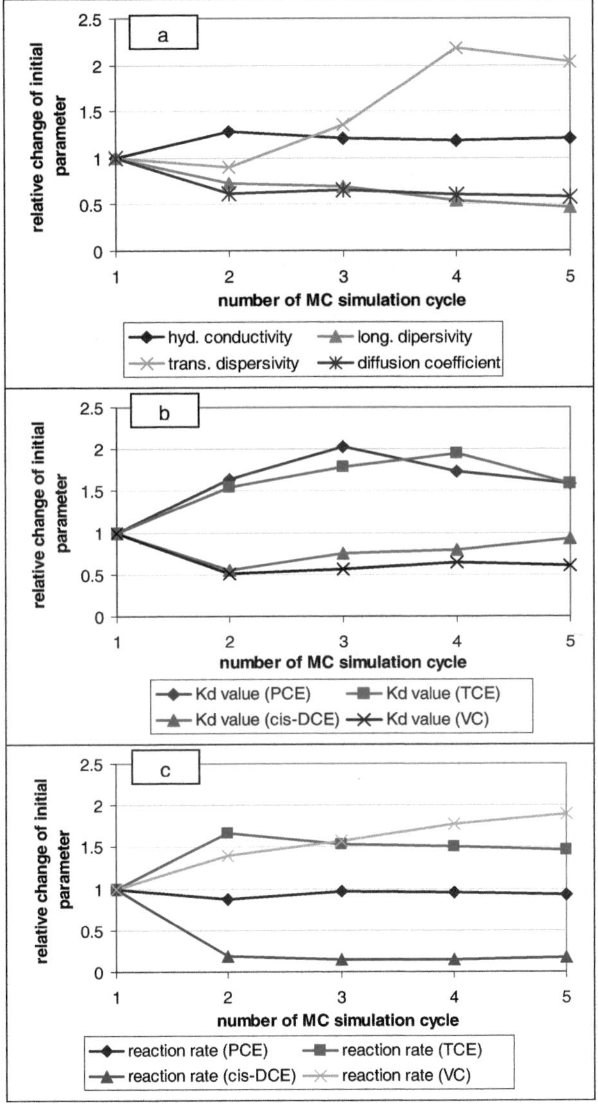

Figure 5.13: Parameter development in the MC simulations normalised to reference model (1: Reference model, 2-5: MC simulations 1-4)

The parameter evolution (shown in Figure 5.13) reveals that some parameters remain almost constant after the first MC simulation (e.g. reaction rates for PCE, TCE and cis-DCE in Figure 5.13c), while other parameters considerably change during the entire optimisation procedure (e.g. sorption coefficients in Figure 5.13b).

Interesting parameter evolutions are observed in case of the degradation rate for VC, which increases by almost a factor of two. However, when comparing the evolution of this parameter with the error development of the VC optimisation (Figure 5.12d) it turns out that this rate seems to have no significant influence on the total error evolution. This observation is in contrary to the outcome of the sensitivity analysis performed in chapter 5.2.1, stating a crucial impact of reaction rates onto model results. It must hence be assumed that VC plume distribution is not improvable with the chosen approach, which is ultimately demonstrated by the error development graph (Figure 5.12d).

Another astonishing observation is the development of longitudinal and transversal dispersivity (Figure 5.12a). In literature data, ratios of 10:1 to 5:1 for both dispersivities are reported. The mathematical optimisation procedure for the longitudinal dispersivity results in at least half of the original value, while the transversal dispersivity almost doubles. The resulting ratio of both is then about 1:1, which is far beyond the numbers found in corresponding literature (Clement et al., 2002; Bordeleau et al., 2008; Benekos et al., 2007). A possible explanation for this behaviour may be the model's implemented steady-state flow boundary condition. In preliminary tests using instationary flow boundary conditions, slightly broader pollutant plumes were calculated with the same set of transport and reaction parameters (see Figure 5.14). Using the steady-state model a similar outcome is achievable by an increase of transversal dispersivity and a decrease of longitudinal dispersivity. This leads to the conclusion that an increased ratio of longitudinal and transversal dispersivity (here approximately 1:1) is able to partially describe plume propagation similar to an instationary flow model (as shown by the parameter evolution during the optimisation approach, Figure 5.13a).

Additionally, from the comparison of instationary and steady-state groundwater models (Figure 5.14) it can be concluded, that the effects of flow instationarities as tested here are small in relation to other transport and reaction parameters, which have proven to contribute to a large extent to pollutant plume propagation (e.g. reaction rate, sorption coefficient etc., see chapter 5.2.1).

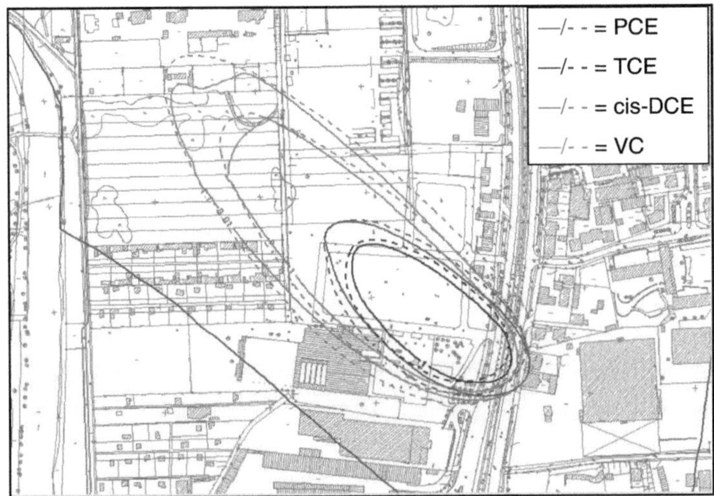

Figure 5.14: Comparison of concentration isolines derived from instationary (dashed lines) and steady-state (solid lines) groundwater model of the Schützenplatz area

Due to small differences in the contaminant distribution of steady-state and instationary model (see Figure 5.14), instationary boundary conditions are neglected during the simulations conducted and described the next chapter. Nevertheless, it has to be denoted that instationarity may still play a crucial role for contaminant propagation in groundwater, but that the model constructed here is obviously not capable to account for this issue. However, more profound considerations and evaluations concerning the influence of oscillating groundwater head as well as on the interaction with bordering ecosystems have to be investigated more deeply.

5.3 Development of a redox-dependant degradation kinetic

5.3.1 Formulation of the problem and conceptual approach

Contaminant degradation in soil is dependant on environmental conditions such as the redox potential (discussed in detail in chapter 2.2.5). As described in chapter 2.1.2 and demonstrated experimentally in chapter 4.2, the reductive dehalogenation reaction of chlorinated ethenes competes with reduction of inorganic electron acceptors, i.e. nitrate and sulphate. After a preliminary reaction model has been developed on the basis of experimental data (chapter 4.2.3), the results are being transferred to higher-scale groundwater models and extended by modifications of the reaction kinetic. The concept developed here is used to improve current understanding of chloroethene degradation in the environment by implementation of naturally occurring processes into the model. Above all, different spatial scales are applied to consider redox-dependant reaction kinetics. A degradation kinetic based on the Monod-equation and further modified with substrate inhibition terms is developed inductively from the experimental results (chapter 4.2) and discussed in chapter 5.3.2.

In chapter 5.3.3 a holistic model for microbially mediated reductive dechlorination is deductively derived from theoretical considerations and subsequently applied to the field studies area "Schützenplatz". In this approach different microorganism species, such as nitrate reducers or sulphate reducers, are implemented according to prevailing redox conditions and contribute to a certain degree to pollutant degradation. To develop and validate a model which is able to successfully deal with these circumstances, a transect model is established to simulate the groundwater flow path on field scale (Figure 5.15). Concentration measurement data reveal that half-way along the transect the contaminant hotspots are found. It is assumed that the pollution source zone is located at the upstream end in the south-east. Therefore, it is supposed that the contaminant plume distribution can be calculated using a two-dimensional model along the transect which comprises the same flow and transport characteristics as the three-dimensional model discussed in chapter 5.1.1.

The observation wells along the transect (B5, B16, B12, SB1, B14, TUBS3 and TUBS1; see Figure 5.15) were sampled several times in order to obtain a comprehensive measurement data set for validation purposes. Along the transect, variable redox conditions were detected: in the upstream region to the south-east nitrate-reducing condition with a redox potential of ~+200 mV were found, while further downstream the iron(III)-reducing conditions with redox potentials of ~-100 mV prevailed.

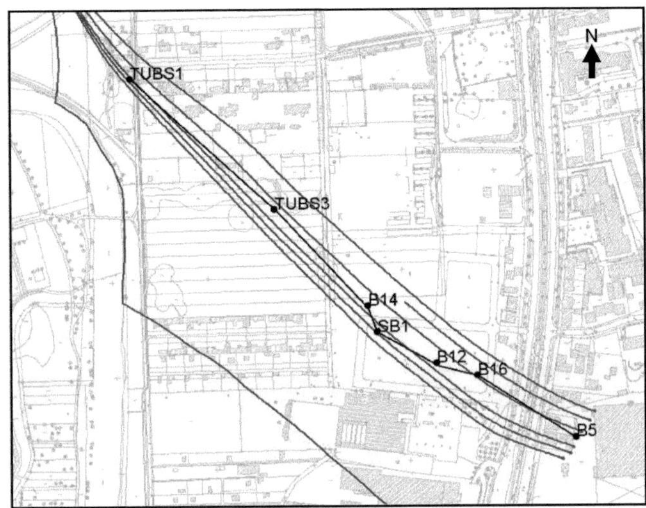

Figure 5.15: Top view of the transect (black line) with observation wells and particle tracking (blue lines)

The results of the concentration measurements are depicted in Figure 5.16. At the upstream end of the transect well B5 is located (referring to 0 m in Figure 5.16), while a distance of 612 m is specified by well TUBS1 in downstream direction. It becomes obvious that a rapid shift in pollutant spectrum from PCE and TCE towards cis-DCE takes place in a region where nitrate is depleted and accumulation of dissolved iron is about to increase. This observation is in good compliance with the experimental results achieved in the batch test experiment (chapter 4.2).

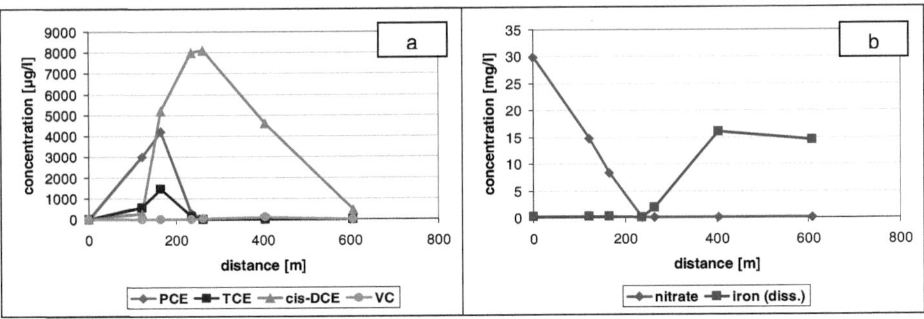

Figure 5.16: concentration distribution of chloroethenes (a) and selected electron acceptors (b) on the transect shown in Figure 5.15

5.3.2 Monod-based model using electron-acceptor inhibition kinetic

The Monod-approach given by equation (2.10) was chosen for a mathematical description of contaminant decay. Modifications of this equation are derived from multiplication with terms describing the influence of inhibiting substances on the reaction of concern (see equation (5.4)). The basic principle of this kinetic approach was evaluated in the batch test experiment (chapter 4.2.2) and the feasibility of this method for degradation modelling was proven via simulation of the experimental results (chapter 4.2.3).

$$r = -r_{max,k}\frac{C_k}{K_{S,k}+C_k}\prod_n\left(\frac{K_{I,n}}{K_{I,n}+C_{I,n}}\right)$$

(5.4)

k: species indicator for chlorinated ethenes

n: species indicator for inhibitors

$K_{I,n}$: inhibition constant of inhibitor n

$C_{I,n}$: concentration of the inhibitor n

In general the above given equation (5.4) enables the consideration of environmental conditions, which influence degradation kinetics of organic pollutants. In detail the reaction equation for TCE are represented in the model as follows (equation (5.5)):

$$r_{TCE} = \eta_{PCE}\cdot r_{max,PCE}\cdot\frac{C_{PCE}}{K_{S,PCE}+C_{PCE}}\cdot\left(\frac{K_{I,NO_3}}{K_{I,NO_3}+C_{NO_3}}\cdot\frac{K_{I,SO_4}}{K_{I,SO}+C_{SO_4}}\right)$$

$$-r_{max,TCE}\cdot\frac{C_{TCE}}{K_{S,TCE}+C_{TCE}}\cdot\left(\frac{K_{I,NO_3}}{K_{I,NO_3}+C_{NO_3}}\cdot\frac{K_{I,SO_4}}{K_{I,SO}+C_{SO_4}}\cdot\frac{K_{I,PCE}}{K_{I,PCE}+C_{PCE}}\right)$$

(5.5)

η_{PCE}: stoichiometric coefficient for formation of PCE

While the upper line of the equation describes formation of TCE due to degradation of PCE, the lower line refers to decay of TCE. Parameterisation of all reaction equations is given in the appendix.

The batch test experiments (see chapter 4.2) revealed that not only nitrate and sulphate can be deemed inhibitive, but also chloroethenes situated in a previous position in the degradation chain may increase the inhibiting effect (e.g. PCE might inhibit TCE degradation to a certain extent). This effect is finally represented by the general equation (5.4).

The approach using this basic inhibition model is implemented directly into the 3D model for field studies and validated according to concentrations in the field (compare chapter 5.1.3). Initial conditions for nitrate and sulphate were chosen according to the depiction in Figure 3.6 and 3.7. The model parameter set for flow and transport characteristics is obtained from the MC optimisation approach (see chapter 5.2.3).

Utilisation of the reaction kinetics given by equation (5.4) delivered far more realistic contaminant plume propagation achieved in the simulation. The quadratic error calculated using this model was 66.5, while the previous simulations using a first-order kinetic had a value of 88.9 even after the MC optimisation approach (Figure 5.17).

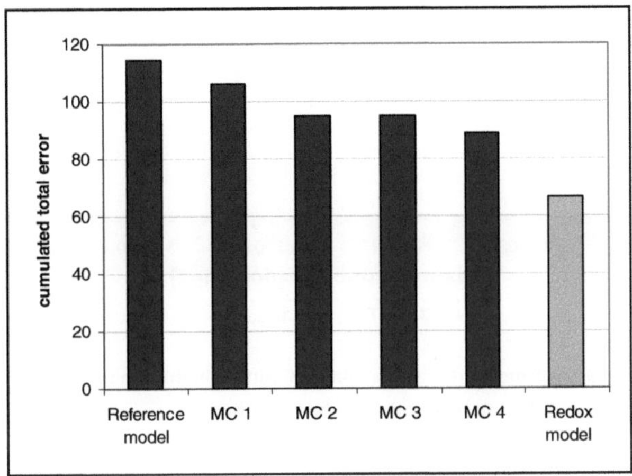

Figure 5.17: Comparison of the cumulated total errors for the model optimisation via MC approach (chapter 5.2.3) and the cumulated total error for the inhibition model

It becomes apparent that the improved reaction kinetic contributes to a major part to the model outcomes, resulting in an almost 25% reduction of the cumulated quadratic error. An additional optimisation similar to the one conducted and described in chapter 5.2.3 might further enhance the simulation results.

An evaluation of simulation results with measurement data is shown in the following figures (Figure 5.18). The comparison with the originally developed reactive transport model (see section 5.1) reveals an improvement of the model. For the concentration scatter plot (Figure 5.18) the increased agreement of measured and simulated data sets is not that apparent (compare Figure 5.6). However, the increased conformity could be demonstrated by calculation of the quadratic deviation (Figure 5.17).

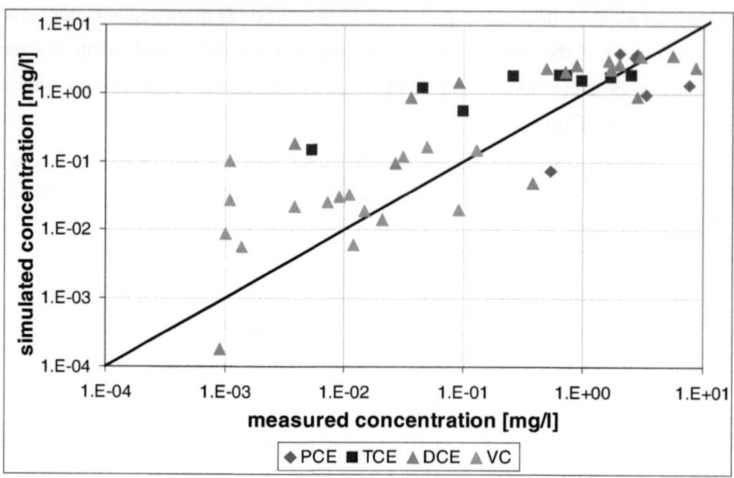

Figure 5.18: Comparison of measured and simulated contaminant concentrations of the redox model referring to chlorinated hydrocarbons

The following graphs 5.18-5.21 show the spatial distribution of contaminants gained from the model approach discussed above.

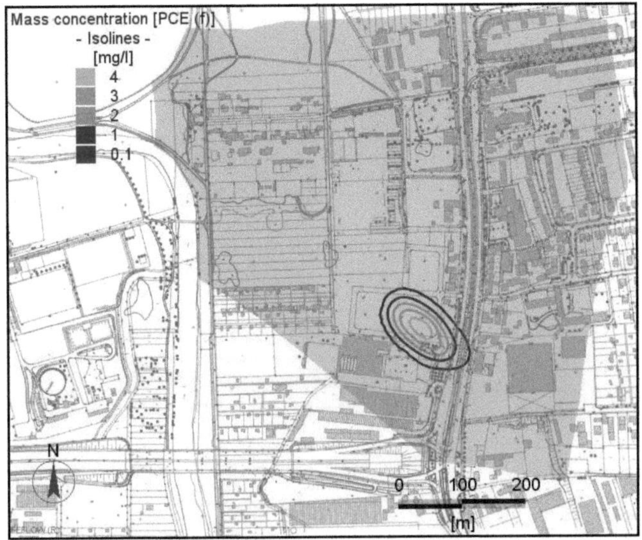

Figure 5.19: Illustration of model results using the simplified redox kinetic applied to the case study area for PCE

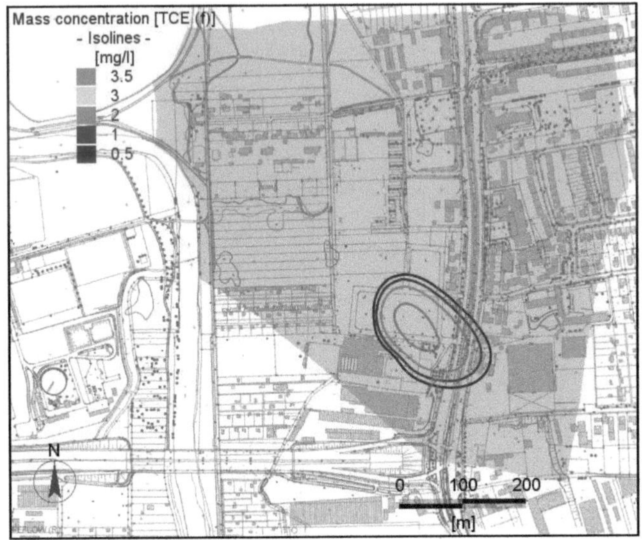

Figure 5.20: Illustration of model results using the simplified redox kinetic applied to the case study area for TCE

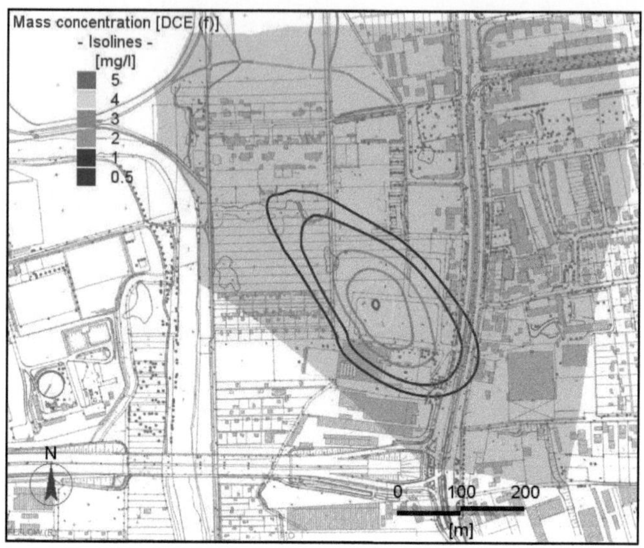

Figure 5.21: Illustration of model results using the simplified redox kinetic applied to the case study area for cis-DCE

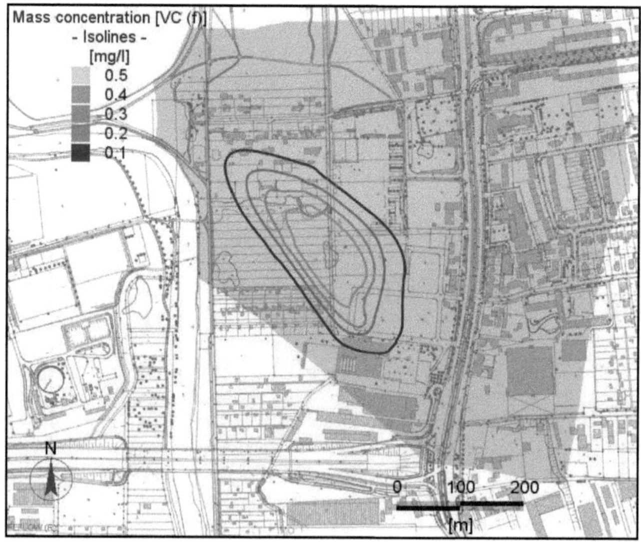

Figure 5.22: Illustration of model results using the simplified redox kinetic applied to the case study area for VC

The isoline illustrations given in Figure 5.19-5.21 make clear that PCE and TCE do not show a distinct tailing of the plume. This fact is also observed in the measured concentrations plot (Figure 3.2 - Figure 3.5). Additionally, cis-DCE is formed quickly after depletion of nitrate (see Figure 3.6). In total a good agreement with measurement data (Figure 3.2-3.5) was demonstrated by using this kinetic model.

5.3.3 Setup of a transect model

On the basis of the aforementioned considerations concerning model simplification (chapter 5.3.1), a two-dimensional flow and transport model comprising a length of 650 m and a width of 50 m is developed. The model is constructed in accordance to the transect along the flow path shown in Figure 5.15 (henceforward called "transect model"). It consists of 16250 rectangular elements (2 m length and 1 m width each) arranged in a single layer. Flow and transport parameters are derived from parameter sets utilised in the preceding simulations using the three-dimensional groundwater models (5.1.1). Preliminary tests reveal an adequate agreement of the transect model with model results from the three-dimensional case study model using the same parameter set. It can thus be assumed that both models deliver comparable results for the observation wells located along the transect. However, computation time is significantly decreased from over three days (3D "Schützenplatz" model) to approximately one hour (2D transect model).

The initial conditions for electron acceptors dissolved oxygen, nitrate, manganese(IV), iron(III) and sulphate are determined by field measurements and interpolated for utilisation in the transect model (see appendix: Figure 11.4). On this basis the electron acceptor distribution is characterised by declining nitrate concentrations (from 35 $mg \cdot l^{-1}$ to 0 $mg \cdot l^{-1}$) and increasing manganese(IV) (from 0 $mg \cdot l^{-1}$ to 5 $mg \cdot l^{-1}$) and iron(III) concentrations (from 0 $mg \cdot l^{-1}$ to 20 $mg \cdot l^{-1}$) to reflect measurement values derived from the case study area (see Figure 5.16b). According to the field measurements, oxygen and sulphate concentrations in the model are kept constant over the entire transect length (0.1 $mg \cdot l^{-1}$ and 400 $mg \cdot l^{-1}$, respectively). All simulations are referred to equal mass of soil organic matter and a corresponding organic carbon content of 0.1%. This value has been determined as an adequate estimate during preceding analyses using soil from a predominant sandy layer found in the test field area.

Further a holistic conceptual approach to model chlorinated ethene degradation was developed. The basic assumption of this model includes the reductive dechlorination of pollutants which is mediated by different soil microorganisms. This model comprises decomposition of soil organic matter $((CH_2O)_n)$ into low-molecular carbohydrates (CH_2O), the metabolisation of these carbohydrates by soil-microorganisms using different electron acceptors (MO_{e-acc}) and finally the degradation of chlorinated ethenes by the consortium of soil microorganisms. Use of this approach enables to describe the natural processes in groundwater under spatially and temporally varying conditions. The conceptual approach is illustrated in Figure 5.23.

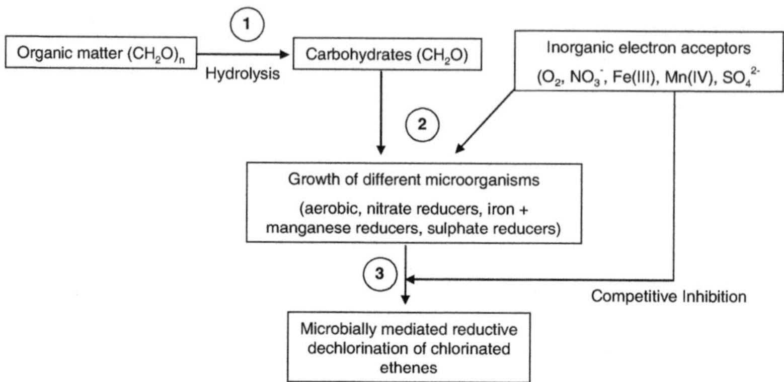

Figure 5.23: Overview of the developed microbially mediated reductive dechlorination model

In the figure above a summary of theoretical background of the comprehensive redox model is depicted. The hydrolysis of ubiquitous soil organic matter into soluble low-molecular carbohydrates (defined as the hypothetical carbohydrate substance CH_2O) is realised by a first-order kinetic approach. Along with inorganic electron acceptors the growth of soil bacteria is mediated. Concentrations of these electron acceptors, such as dissolved oxygen or nitrate, were determined during field measurements. The different bacteria catalyse to a certain extent the decomposition of chlorinated ethenes, but are otherwise inhibited by the different inorganic electron acceptors.

It has to be denoted that due to the high complexity of the underlying processes an exact proof of each reaction pathway cannot be performed. The involved soil bacteria and their mass can only be estimated. For validation purposes the same technique as described in chapter 5.1.3 is utilised. However, it still needs to be emphasised that this method is a conceptual approach and mainly comprises theoretical considerations, which are applied to naturally occurring processes and reaction pathways.

Related to the biological degradation processes the mathematical formulation is performed and is drawn on the following equations:

1. Hydrolysis of insoluble carbohydrates of the soil organic matter (defined by the hypothetical chemical formula $(CH_2O)_n$) and respective formation of soluble carbohydrates (given by the formula CH_2O) according to a kinetic reaction equation of first-order:

$$v = -k_v \cdot C_{SOM} \tag{5.6}$$

    k_v: reaction rate constant of soil organic matter decomposition [s^{-1}]

    C_{SOM}: concentration of soil organic matter [$mg \cdot l^{-1}$]

2. Growth of different microorganisms utilising electron acceptors is described by an extended Monod-equation. In this approach two limiting substrates (carbohydrate and respective electron acceptor) determine growth of each microorganism, while a death rate limits maximum concentrations and is realised by a first-order equation:

$$\mu = \mu_{max,j} \cdot \frac{C_{CH2O}}{K_{S,C,j} + C_{CH2O}} \cdot \frac{C_{e-acc}}{K_{S,e,j} + C_{e-acc}} - k_{D,j} \cdot C_{MO,j} \tag{5.7}$$

    $k_{D,j}$: death rate of microorganism species j

    $C_{MO,j}$: concentration of microorganism species j

    $\mu_{max,j}$: maximum growth rate of microorganism species j

    $K_{S,C,j}$: half-saturation constant of microorganism species j and carbohydrate-substrate

    $K_{S,e,j}$: half-saturation constant of microorganism species j and electron-acceptor substrate

    C_{CH2O}: concentration of soluble carbohydrates

    C_{e-acc}: concentration of specific electron acceptor

3. Reductive dechlorination of chloroethenes follows an extended Monod-approach similar to the method presented in the previous chapter 5.3.2. Inhibition of dechlorination also works in the same way (e.g. see also equation (5.4)). Additionally, the different microorganisms mediating degradation of chlorinated ethenes contribute to the degradation kinetic with a specific factor.

$$r = \sum_k \left(C_{MO,j} \cdot r_{max,m} \cdot \frac{C_{CHC,i}}{K_{M,CHC} + C_{CHC,i}} \cdot \left(\prod_k \frac{I_{e-acc,k}}{K_{I,e-acc} + I_{e-acc,k}} \right) \right) \tag{5.8}$$

    $r_{max,m}$: maximum degradation rate of chloroethene m

    $I_{e-acc,k}$: concentration of inhibiting electron acceptor

    $K_{I,e-acc}$: inhibition constant of electron acceptor k

    $C_{MO,j}$: concentration of dechlorinating species j

Parameter sets for the simulations can be found in the appendix (chapter 11.9).

The results in Figure 5.24 for the pollutant concentration development in the transect model refer to 50 years simulation time. It becomes obvious that the model predicts the real concentrations adequately with small over- (e.g. PCE) or underestimations (e.g. VC). However, in view of the specified high complexity of the underlying reaction system and the uncertainties in measurement field data, the outcome is satisfactory. This model approach based on a microbially-mediated reductive dechlorination kinetic obviously produces reliable results at least for the simplified two-dimensional transect model.

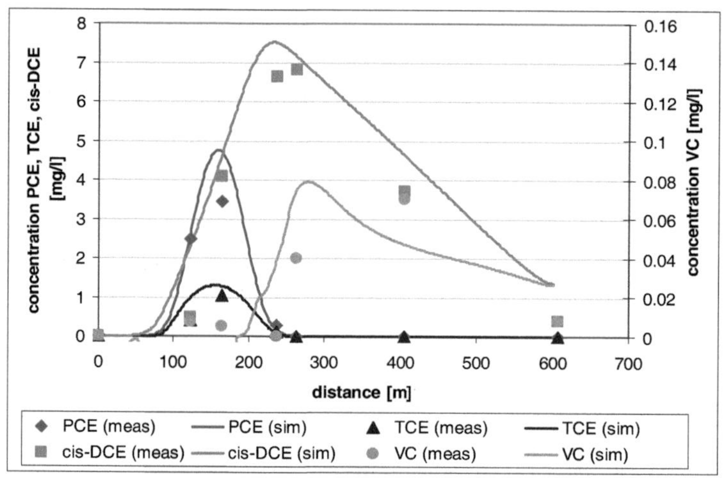

Figure 5.24: concentration development of the transect model using the holistic reductive dechlorination approach

Based on the results shown in Figure 5.24, a full-scale 3D model was parameterised with the equivalent parameter set and reaction equations as used in the transect model. The additionally necessary properties such as head boundary conditions are taken from the model optimisation approach described in chapter 5.2.3. Inverse distance interpolation of the mean measurement values (determined during several measurement campaigns) served as initial conditions for electron acceptor concentrations (e.g. see Figure 3.6, Figure 3.7).

The outcome of this model is evaluated in analogy to the transect model: concentrations are compared on the basis of measured concentrations for the observation wells (B5, B16, B12, SB1, B14, TUBS3 and TUBS1). The results fit quite well for the compounds PCE and TCE, whereas cis-DCE and VC are slightly underestimated by the model (see Figure 5.25).

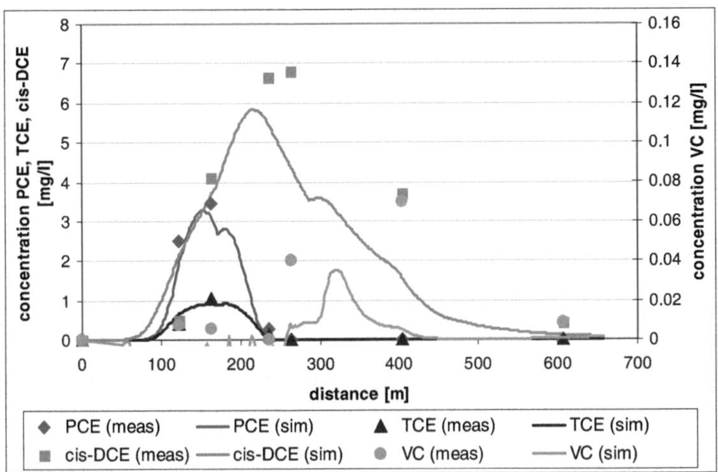

Figure 5.25: Comparison of the results for the model scale-up towards the case study model: parameter set of the 2D transect model implemented into the 3D case study model

The good agreement of the transect (Figure 5.24) and full-scale model (Figure 5.25) is taken as a proof for the feasibility and comparability of the 2D transect model with the 3D full-scale model. The lower concentrations of the 3D model may be traced back to dilution effects caused by transversal dispersivity, which does not play a significant role in the 2D model. However, the small range of differences between both models make the advance conducted here promising for further investigations dealing with scale-up improvement.

Nevertheless, it must be denoted that the computational effort required for the implementation of detailed degradation models like the one presented here is a drawback of such approaches. Thus, for the development and evaluation of the health risk model (chapter 6), the less complex inhibition model described in chapter 5.3.2 is applied due to significantly lower hardware requirements and calculation time.

6 Risk assessment approach

The risk assessment approach utilised here emanates from a health risk approach which additionally integrates the reactive transport model described previously. This approach intertwines both, enhanced groundwater modelling and mathematical derivatives to calculate and evaluate risk including uncertainties for health hazards. The risk formulations published by the U.S. EPA are adopted for a calculation of health risk posed by groundwater contaminations on the regional scale. As described earlier, risk models are often applied to prove an underlying model concept based on hypothetical scenarios (Benekos et al., 2007) or rather simple groundwater models (Kaufman et al., 2009; Massabo et al., 2008). Generally, those models are focused on one aspect of the multidisciplinary characteristic which is either groundwater modelling, risk assessment or uncertainty calculation. The need to combine these multidisciplinary aspects in a holistic way is apparent and hence, is performed in this chapter. The application of the risk evaluation framework towards a real world scenario aims to assist in decision making processes in the course of remediation measures.

6.1 Background and model assumptions

The evaluation of the health risk is performed according to the approach discussed in chapter 2.3. Further, general assumptions are made (see below), to describe and simulate risk scenarios close to reality. Reference to a real world scenario is established by the case study area and field data acquisition (chapter 3).

The calculation of the average daily dose is based on equation (2.18) with some modifications. The modifications concern the time and concentration averaging factors (AT, ED) which can be omitted in the simulation as concentration values are calculated continuously over the entire model time scale. Thus, the equation is modified and can be formulated to calculate the daily dose (DD_k) as follows:

$$DD_k = \overline{C}_{kd}\left(\frac{GR}{BW}\right) \cdot EF \tag{6.1}$$

$\overline{C}_{kd}$: average daily concentration of contaminant k in tap water [mg·l^{-1}]

GR: ingestion rate of tap water [l·d^{-1}]

BW: body weight [kg]

EF: exposure frequency [d·y^{-1}]

The groundwater model applied here is based on the following assumptions:

- Boundary conditions were chosen according to the values discussed in the previous chapter, which can also be found in the appendix (chapter 11.8).
- Steady-state flow conditions were used according to the calibrated flow model explained in chapter 5.1.2. Instationary flow conditions are considered as well, but model runs revealed only small differences between concentrations obtained from steady-state and instationary flow conditions (compare with Figure 5.14).
- Flow and transport parameters as well as reaction equations were selected according to the Monod-model using electron-acceptor inhibition kinetics as presented in chapter 5.3.2.
- The simulated contamination plume refers to a simulation time of 50 years. Thus, it is assumed that contaminant release started in 1960 and lasted until 1975. Onward from this date, pollutant infiltration stops and contaminant fate is hence only subject to transport and degradation.

For evaluation of the health risk additional assumptions are required and specified as follows:

- Risk assessment refers to the residential area adjoining the test area in the north-western part of the domain (see Figure 6.1). As a matter of fact no municipal drinking water supply was installed during the years 1960-2010 (according to personal communication with residents, municipal water supply was installed in summer 2010). Hence, each house is supposed to have its own groundwater well.
- As a worst-case scenario it is assumed that filtered, but furthermore untreated groundwater is used for drinking, cooking and other household purposes. This groundwater is supposed to contain a certain amount of contaminants as computed by the groundwater model. The total oral daily ingestion (GR as described by equation (2.18)) is estimated to be normal-distributed with a mean value of 2 $l \cdot d^{-1}$ along with a standard deviation of 0.5 $l \cdot d^{-1}$.
- Population inhomogeneity is also represented by a Gaussian distribution with a mean bodyweight (BW) of 75 kg and a standard deviation of 12.5 kg.
- The exposure frequency (EF) is considered to be normal-distributed with 335 $d \cdot y^{-1}$ as mean value and 15 $d \cdot y^{-1}$ as standard deviation.
- The inherent health risk is calculated in the two ways represented by equation (2.17) for a cancer-risk approach and equation (2.19) for the Hazard Quotient approach. Respective parameters implemented for the risk assessment of oral exposure are given in Table 2.7.

Given these considerations a varying vulnerability of individuals of the respective population in terms of contaminant impact is specified. High vulnerability to toxic or carcinogenic substances thus results from low body-weight, high oral ingestion rate and a high exposure frequency. For simplification purposes the population inhomogeneity is described by the 10% quantile (referring to the 10% part of least vulnerable persons), the median value (corresponding to the 50% quantile) and a 90% quantile (all residents except for the 10% most vulnerable residents).

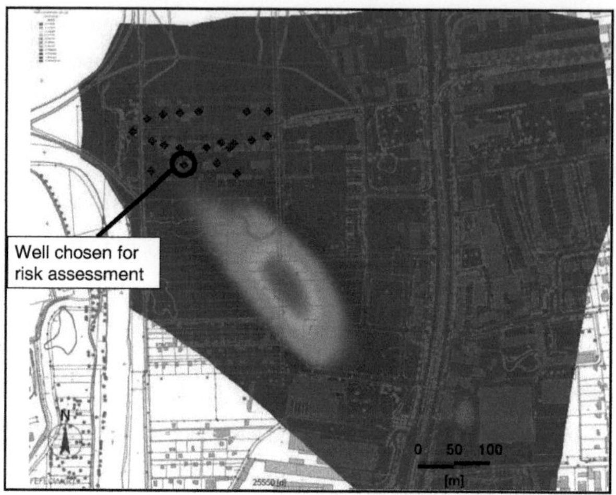

Figure 6.1: Overview of the model domain with location of contaminant source area (in red, south-east), assumed residents' groundwater wells (blue dots, north-west) and calculated cis-DCE distribution to illustrate the intrinsic risk situation

The performed risk assessment, which is based on the previously presented assumptions, is understood as an upper bound estimate for the impact on the affected population. The risk evaluation approach is divided into two different uncertainty assessments: First, the aquifer uncertainty calculated with MC methods (compare chapter 5.2.2), and second, the population variability which is taken into consideration for a population cohort consisting of different individuals. The parameters discussed above were generated using the Mersenne twister algorithm (as implemented in Mathworks Matlab 2009b).

Despite the fact that municipal drinking water supply was installed in 2010, risk calculation is also simulated for the next 20 years (until 2030) and hence totally refers to a simulation time of 70 years (from 1960 to 2030). This simulation time is the same used for calculating probabilities of contaminant occurrence as described in chapter 5.2.2.

6.2 Modelling health risk

6.2.1 Cancer risk approach

Based on the assumptions made in the previous chapter another MC simulation was performed integrating the redox-dependant reaction kinetic discussed in chapter 5.3.2. The previous MC simulation (described in chapter 5.2.2) yielded that 100 model variants are considered sufficient in terms of stochastic variance (see Figure 5.8). Therefore the same number of variants was chosen for the approach discussed here. The groundwater model described in chapter 5.3.2 was utilised as reference model for the MC simulation. Flow, transport and reaction parameters discussed in chapter 5.2.2 were additionally varied to account for aquifer uncertainty. Further, population variability was implemented according to the model assumptions described in chapter 6.1.

Due to the impossibility of depicting temporal and spatial outcomes in the same graph, the analysis of the simulation results is performed in two different ways:

a) The temporal evolution of the cumulated cancer risk according to equation (2.17) is conducted for an exemplarily chosen observation well in the study area. The model outcomes are shown in Figure 6.2 and Figure 6.3. Besides the median risk evolution, 10% and 90% quantiles for the model uncertainty as well as the for the population inhomogeneity are displayed. By use of these quantiles the influence of outliers on the calculated results is decreased and the impact of both, model uncertainty and population variance can be assessed.

b) The spatial analysis of risk was conducted in analogy to the procedure described in chapter 5.2.2 for 50 years of simulation time, referring to the present time. Again, model uncertainty and population inhomogeneity are included in the analysis: in Figure 6.4a the model uncertainty is depicted in terms of probability isolines (for 50%, 80% and 90% exceedance of a 0.001 risk level). For population variance the median value as well as the 10% and 90% quantiles are shown in Figure 6.4b.

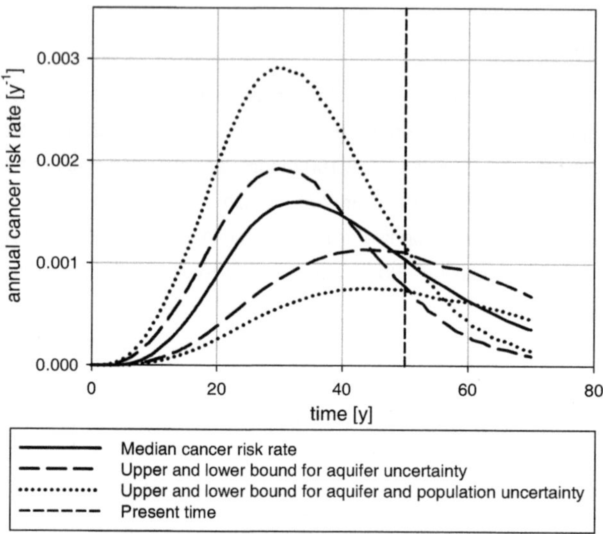

Figure 6.2: Cancer risk rate development with upper and lower bounds for transport model uncertainty and population variability exemplarily shown for one groundwater well in the case study area (see Figure 6.1)

a) Figure 6.2 demonstrates the annual cancer risk rate for the example groundwater well, which was chosen for risk assessment. Maximum cancer risk rates are calculated for a simulation time of 30 to 50 years, referring to the years 1990 until 2010. Additionally it becomes apparent, that lower cancer risk rates come along with a longer exposure time. This effect is explainable by the properties of the reactive transport model: due to the fact that approximately the same amount of pollutants is released into the environment, a quicker transport of contaminants automatically leads to an earlier and more intense occurrence at the risk receptor site in comparison to a more retarded groundwater transport.

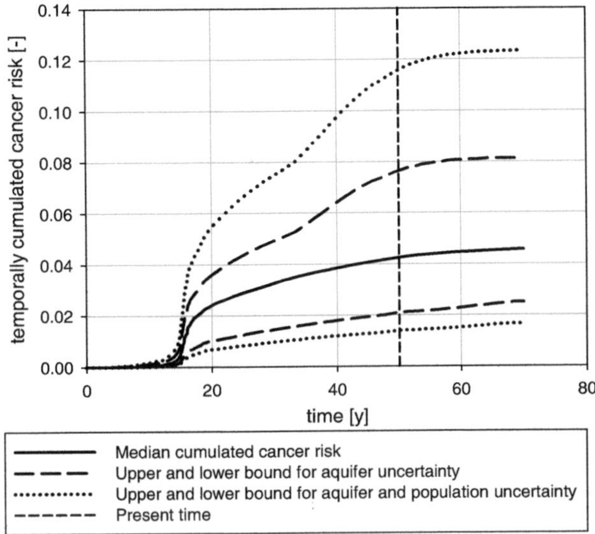

Figure 6.3: cumulated cancer risk evolution with upper and lower bounds for model and population uncertainty (10% and 90% quantile, respectively) exemplarily shown for one observation well in the study area (see Figure 6.1)

Figure 6.3 reveals a strong increase of cancer risk within 10 to 20 years after the initial simulation time. This outcome corresponds with the observation that contaminant migration in groundwater towards the downstream region of the study area (where the well is located) takes several years. After a strong initial rise of cancer risk the subsequent increase slows down. This may be explained with depletion of the contaminant source after 15 years of simulation time and due to degradation processes; lower contaminant concentrations subsequently lead to flattened slope of the cancer risk curve.

It becomes obvious that groundwater model uncertainty leads to a factor of approximately ±2 in the risk outcomes and that population uncertainty additionally contributes to cancer risk with about the same factor. In total, the combination of the two uncertainty assessments results in a deviation of cancer risk outcomes by almost factor 10.

Risk evaluation of single compounds (Table 6.1) shows that VC predominantly contributes to the evolution of cumulated cancer risk. This result can be explained by two reasons: First, VC is mainly found in the area of the considered example well. At this location the higher substituted compounds, i.e. PCE and TCE, are already degraded due to the shift in redox conditions as implemented in the groundwater model. Second, VC has a considerably high cancer slope factor (see Table 2.7). In comparison to the other compounds (CSF_{PCE}: 0.01-0.1, CSF_{TCE}: 0.05) the CSF of VC is almost ten times higher (CSF_{VC}: 0.72). This results in the observation that even at small VC concentrations a significant increase in cancer risk occurs.

Cis-DCE is not accounted for the assessment of cancer risk due to the non-existent cancer slope factor (U.S. Environmental Protection Agency, 2010).

Table 6.1: Contribution of single chemical compounds to the cancer risk at the example well

Compound	Contribution to cancer risk
PCE	~0
TCE	~0
cis-DCE	n.a.
VC	~100%

b) Figure 6.4 summarises the evaluation of the spatial extent of risk for all chloroethene compounds. The depicted risk threshold of 0.001 was chosen arbitrarily. Figure 6.4a shows the groundwater model's variance in terms of probability isolines for the cumulated cancer risk of all chloroethenes. The solid line represents a 50% probability of exceedance of a 0.001 risk threshold, which refers to the fact that 50 (of a total of 100) model variants of the MC simulation exceed the given risk level within the plotted isoline. The dashed lines indicate exceedance of 80% and 90% probability, respectively.

Figure 6.4b depicts the variability of the affected population. The isolines were generated on the basis of the 80% probability isolines for the model uncertainty. Here the solid line accounts for the 90% population quantile, which refers to 90% least vulnerable proportion of the population. The dashed lines represent the median quantile and the 10% quantile of the population, i.e. the 50% and 10% least vulnerable population part.

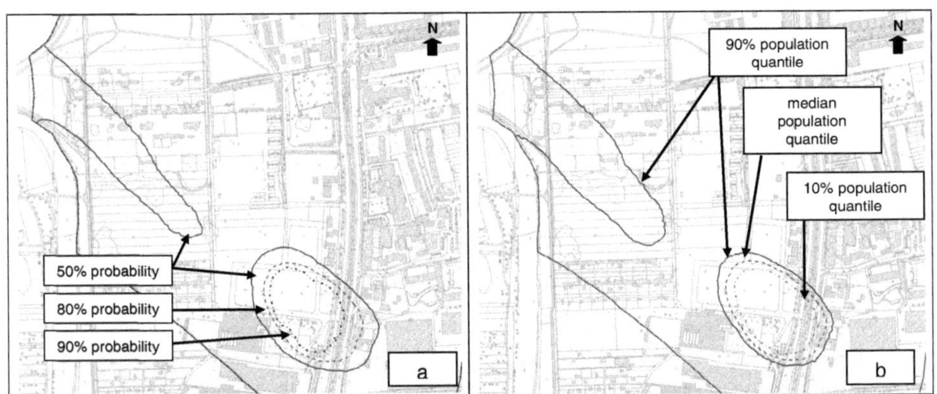

Figure 6.4: Cancer risk for current time period (50 years simulation time); a: probability distribution for exceedance of a risk level of 0.001 cumulated over all chloroethene compounds for the population median; b: uncertainty evaluation using fixed aquifer model uncertainty and varying population quantiles at a risk level of 0.001

As shown in Figure 6.2 – Figure 6.4, uncertainty and variability in the risk assessment procedure of groundwater contaminations can be represented using temporal and spatial depiction. The analysis for a single groundwater well exemplarily delivered the estimated direct impact onto the residents at risk (Figure 6.2 – Figure 6.3). It could be shown that cancer risk in the downstream part of the study area predominantly originates from the presence of VC in groundwater (Table 6.1). However, when looking at the spatial distribution of cancer risk, much higher values can be found in the pollution's core area further upstream (Figure 6.4). Additionally, the fact that cis-DCE does not pose a cancer risk leads to a dissection of affected areas into the upstream part, where PCE and TCE are found in the aquifer, and the downstream part, where mainly VC is responsible for cancer risk.

The calculated risk values indicate a ~2-12% increased cancer risk over the total simulation time of 50 years for inhabitants in the affected area given the assumptions made in this assessment. Again, it has to be emphasised that these outcomes are considered as worst-case scenario. However, in comparison to other risk assessments found in literature for volatile organic compounds (Fan et al., 2009; Khadam and Kaluarachchi, 2003), these values are comparably high, leading to an urgent need of remedial measures. Additionally, the cancer risk assessment clearly proves that utilisation of contaminated groundwater for household purposes is not recommendable in the contaminated area. Regulatory actions are thus advisable to protect the population.

6.2.2 Hazard Quotient approach

The non-cancer related risk approach is based on the Hazard Quotient discussed in detail in chapter 2.3.3. It is computed according to equation (2.19) using respective reference doses for each chloroethene compound as given in Table 2.7. Again, the modified daily dose DD_k as formulated in equation (6.1) is applied for the calculation. The resulting Hazard Quotient describes the ratio of an ingested daily concentration compared to a specific concentration of the respective compound which is regarded as non-hazardous.

In similarity to the cancer risk calculation in the previous chapter, risk evolution is assessed in a temporal (a) and spatial (b) manner. The result of the temporally resolved Hazard Quotient is depicted in Figure 6.5.

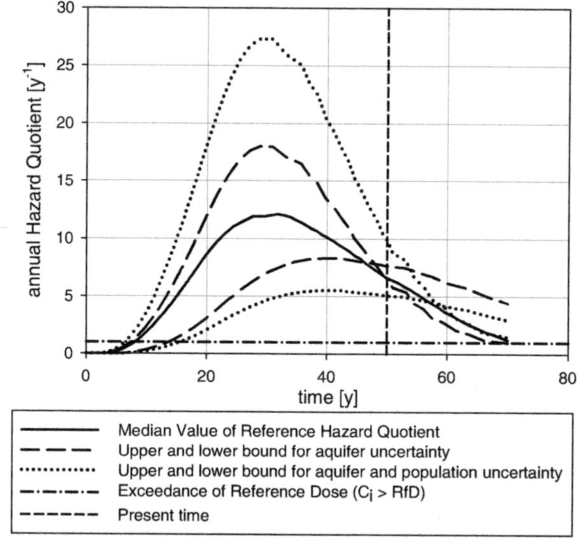

Figure 6.5: Temporal evolution of median annual Hazard Quotient according to equation (2.19) with upper and lower bounds for model and population uncertainty (10% and 90% quantile) for an example well in the study area

a) Figure 6.5 exhibits the trend for the annual Hazard Quotient for the example observation well. It becomes obvious that in all depicted model variants the Hazard Quotient exceeds the value 1, which means that a potentially hazardous effect on human health cannot be excluded. Another observation derived from the graph is the fact that lower Hazard Quotients come along with long-lasting impact, while high absolute Hazard Quotients often result in a faster decline of risk. This result is similar to the cancer risk approach (Figure 6.2).

In Table 6.2 the single compounds' contribution to the Hazard Quotient at the example well is given. In contrast to the cancer risk assessment, cis-DCE plays a crucial role in the outcomes of the Hazard Quotient calculation. High concentrations of cis-DCE in the downstream part of the area along with the comparably low reference dose for this compound ($RfD_{cis-DCE}$: 0.002 mg·kg^{-1}·d^{-1}) are the reason for this result. The influence of VC on the Hazard Quotient is much smaller and is explained by low occurring concentrations. PCE and TCE do not have any impact onto the risk value at the example well due to the degradation processes which reduce respective concentrations to zero before these compounds reach the example observation well.

Table 6.2: Contribution of single chemical compounds to the Hazard Quotient at the example well

Compound	Contribution to Hazard Quotient
PCE	~0
TCE	~0
cis-DCE	~94.3%
VC	~5.6%

b) The spatial analysis of the non-cancer risk was performed separately for each chloroethene. Results for this calculation are shown in Figure 6.6 for model uncertainty.

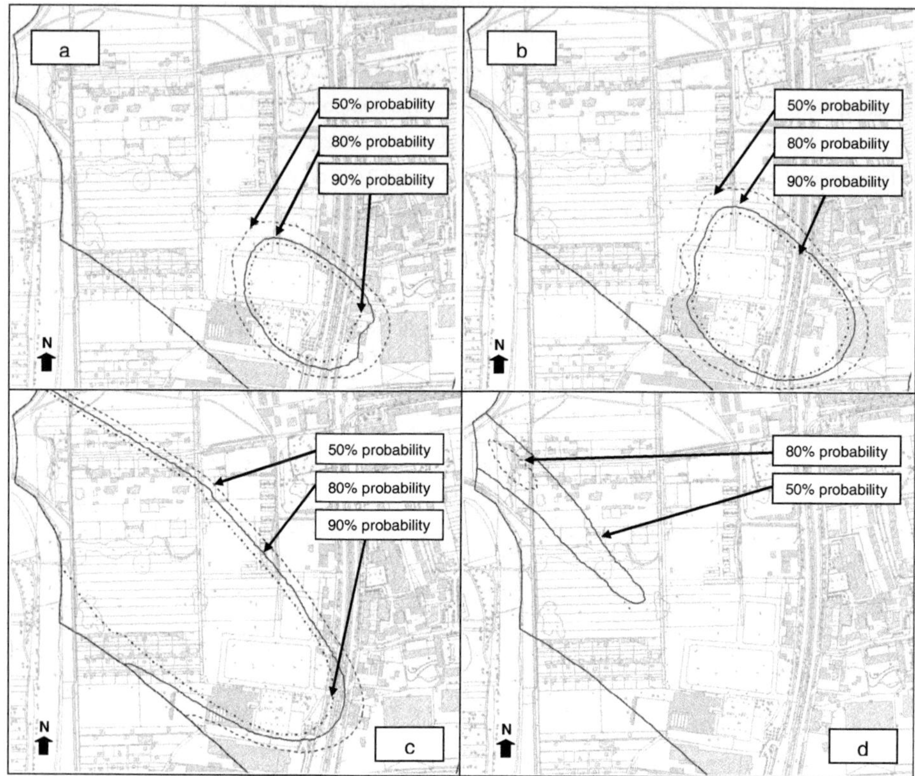

Figure 6.6: Exceedance of Hazard Quotient (HQ > 1 according to equation (2.19)) at 50 years simulation time with regard to model uncertainty; a) PCE, b) TCE, c) cis-DCE, d) VC

Figure 6.7 depicts the spatial outcomes of the models with respects to population inhomogeneity.

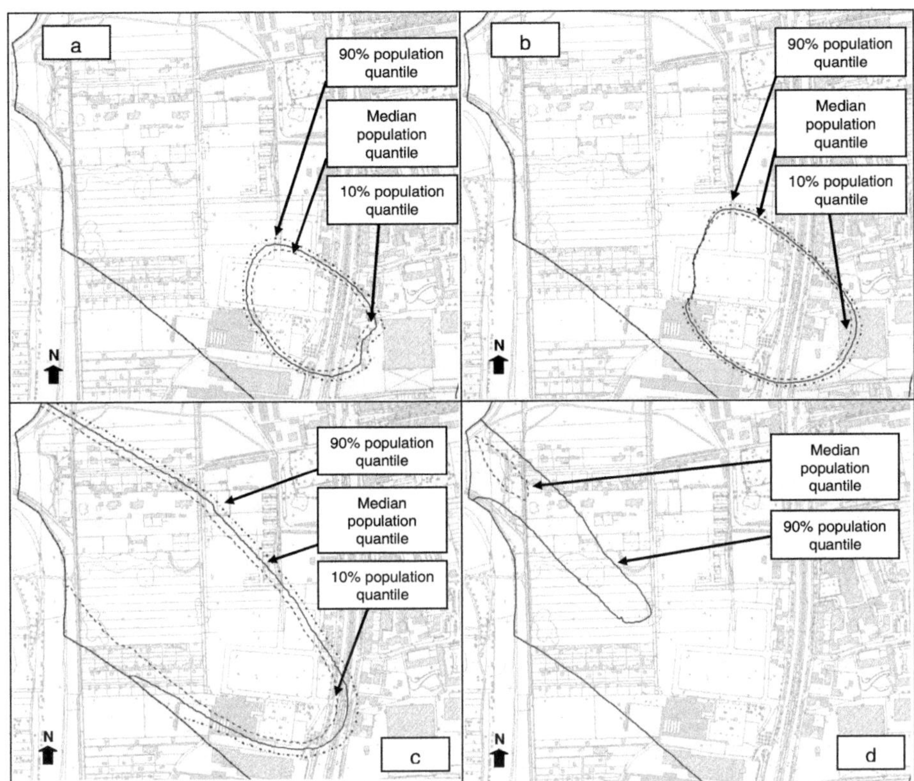

Figure 6.7: Exceedance of Hazard Quotient for each chloroethene compound (at 50 years simulation time) with regard to population uncertainties, a) PCE, b) TCE, c) cis-DCE, d) VC

The comparison of uncertainty resulting from both, groundwater model uncertainty and population variability, shows similar results concerning the isoline distribution. It becomes obvious that the threshold Hazard Quotient of 1 is exceeded in large areas of the domain. For most of the simulated chloroethenes (e.g. PCE, TCE and cis-DCE in Figure 6.7) the uncertainty assessment does not exhibit major differences with respect to the isoline distribution, i.e. probability isolines do not differ much from each other. This observation is caused by the distinct exceedance of the threshold value in all model variants (see Figure 6.5).

On the basis of the reference doses given in Table 2.7, the daily amount for a resident with median vulnerability is almost 180 µg for PCE, 20 µg for TCE, 90 µg for cis-DCE and 130 µg for VC. These concentrations are partially exceeded by the uptake of a very small amount of contaminated groundwater (e.g. approximately 10 ml groundwater with a concentration of 2000 $\mu g \cdot l^{-1}$ TCE). In case of a mean daily ingestion rate of 2 l groundwater per day as given by the model assumptions, these values are exceeded by concentrations of 90 $\mu g \cdot l^{-1}$ for PCE, 10 $\mu g \cdot l^{-1}$ for TCE, 45 $\mu g \cdot l^{-1}$ for cis-DCE or 65 $\mu g \cdot l^{-1}$ for VC, respectively. For the case study example, simulated concentrations for chloroethenes range up to several thousand $\mu g \cdot l^{-1}$, which indicates by all means an inherent danger to the public using groundwater for household purposes: chlorinated ethenes in the affected area might lead to potential health impacts in the study area.

7 Summary and discussion

The primary goal of this work is the establishment and application of a risk assessment for groundwater contaminations on the field scale. This aim requires a precise prediction of contaminant occurrence and the evolution of the pollutant plume extent. Modelling techniques have proven to be reliable for this purpose. However, currently a lack of knowledge regarding the interdependencies of flow, transport and reaction processes in the aquifer exists, leading to ambiguity in the predicted contaminant concentrations. Additionally, uncertainties in aquifer parameters resulting from a limited number of groundwater wells are difficult to take into account. To overcome these difficulties, the here presented work combines and concatenates methods from different scientific disciplines such as microbiology and hydrogeological modelling. On the basis of laboratory scale experiments and field measurements, an extended and refined risk formulation has been derived from a comprehensive reactive transport model and contaminant-specific probabilities of occurrence. In a further step the risk model is applied to a real-world scenario to prove the feasibility of the model approach. Based on model results several conclusions concerning the complex interactions of biologically mediated contaminant degradation in an aquifer system can be drawn, e.g. the focus on environmental conditions like redox potential during site survey measures.

In general, the approach to start with small-scale laboratory experiments for parameter estimation (e.g. transport parameters in chapter 4.1, reaction parameters in chapter 4.2) turned out to be quite reliable. It was shown that experimentally determined parameters using soil and groundwater from the test area are in good accordance with literature data (e.g. Wiedemeier et al., 1998). Additionally, the inhibiting mechanism of electron acceptors towards reductive dechlorination was demonstrated in a batch experiment and could confirm the corresponding model approach. Simulations of both, soil column (chapter 4.1) and small batch-reactor experiments (chapter 4.2), have proven to be successful and yielded valuable information about inhibition parameters, which were implemented in the groundwater model.

The microbiological screening of chloroethene-degrading bacteria (chapter 4.3) was used to link dechlorinating bacteria, prevailing redox conditions, and different contaminant species. Indeed a certain group of bacteria capable of reductive dechlorination was detected in the groundwater samples (strain DF-1 and o-17 according to Watts et al., 2005). Unfortunately, no relation between redox conditions and the presence of corresponding microorganisms was found. Additionally, the most important strain for complete mineralisation of chloroethenes, *Dehalococcoides ethenogenes* 195, could not be detected in any of the samples. These results do not necessarily imply the absence of the respective strain or that cis-DCE will not be degraded in the long term, but obviously environmental conditions in the test area (e.g. high sulphate concentrations) do not favour degradation beyond cis-DCE.

In a further step the results obtained from the preliminary experiments were concatenated with initial conditions determined during field site measurement campaigns (chapter 3.2) and merged into a three-dimensional flow, transport and reaction model to assess contaminant migration in groundwater. This model was validated against field site data regarding pollutant concentrations and served as initial model for subsequently developed groundwater models (chapter 5.1). Nevertheless it became obvious, that the spatial extent of the pollutant plume in an area of 0.4 km^2 is difficult to assess with data from only around 20 observation wells. In addition, concentration data in the wells partially exhibited a fluctuating behaviour. Especially in the peripheral zones of the plume and in the downstream region of the study site, well density is rather low and thus, informative value of the measurement data regarding spatial extent of the contamination is improvable by additional observation wells. Future groundwater monitoring campaigns should therefore focus on an even and well-directed distribution of observation wells throughout the contaminated area of interest to provide a reliable status of the pollutant plume. Additionally, establishment of a groundwater model using instationary flow boundary conditions as applied in the doctoral thesis of Kathrin Helmholz may yield more accurate predictions about contaminant plume propagation.

However, even with highly accurate information concerning the plume extent, the most complex groundwater model is not able to correctly deal with inhomogeneities in soil and groundwater flow. Besides this the calculation of dispersivity often causes problems in groundwater modelling. Due to the wide-spread implementation of the Bear-Scheidegger formulation for dispersion effects which assumes identical values for transverse and vertical dispersion (Diersch, 2009), the vertical dispersivity is not described appropriately. Literature regarding this issue reports far smaller values for vertical dispersivity (Garabedian et al., 1991; Robbins et al., 1989). Modelling depth-resolved concentration data is therefore limited and hence, contaminant concentrations are implemented and evaluated on a two-dimensional basis in this work.

Nevertheless, a sensitivity analysis of important flow, transport and reaction parameters, followed by an uncertainty assessment based on the MC technique, was utilised to evaluate the influence of flow, transport and reaction parameter variation (chapter 5.2.1 and 5.2.2). In order to further improve the model outcomes an optimisation of these parameters was conducted using a MC approach (chapter 5.2.3). Related to PCE and TCE acceptable results were achieved using this method. It became obvious that mathematical parameter optimisation does solve problems originating from parameter uncertainty, but not other model-inherent uncertainties (e.g. those resulting from boundary conditions). Additionally, the need for improvement of the underlying reaction kinetics became obvious throughout the MC optimisation and was performed subsequently.

An extended degradation kinetic was developed, which yielded useful information for the establishment of precise models dealing with contaminant degradation. This development was subdivided into two different approaches:

1) On the basis of field measurement data and parameters derived from the batch test experiment, an inhibition of reductive dechlorination by nitrate and sulphate was formulated and applied to the field-scale model (chapter 5.3.2).

2) Additionally, a model deducted from pollutant degradation experiments was established. It was necessary to simplify the underlying transport model towards a transect model due to the high complexity and computation time of this approach (chapter 5.3.3). Competitive results were obtained using this model and even up-scaling of the kinetic approach on the field-scale model worked well, but the increase in accuracy of model results in comparison to the inhibition model did not justify the computational effort required. The necessary calculation time of around three days on a standard desktop computer (quad-core processor, 2.5 GHz, 4 GB RAM) restricted further application of this model. Therefore, the inhibition model was utilised for subsequent health risk calculations.

The next task of this work was the development of a risk model to assess the health risk of different chemicals and, as prove of concept, its application to a field-site area. The health risk calculation published by the U.S. EPA was used for the case of oral ingestion of contaminated groundwater. Although these circumstances may be seen as worst-case scenario, uptake of groundwater is still imaginable due to the fact that municipal water supply in the area was installed only recently.

Two components were merged together in order to assess cancer (chapter 6.2.1) and non-cancer related (chapter 6.2.2) health risks for residents of the affected model area: based on the reactive transport model developed in this work, the predicted concentration data were used to describe the exposure to toxic and/or carcinogenic substances. The uncertainty assessment was performed by means of a MC simulation for the contaminant transport model as described in chapter 5.2.2 (Greis et al., 2011). Further, residents of the study area serve as risk receptors with a variable vulnerability that is caused by population inhomogeneities. Hence, risk was evaluated in a probabilistic approach including a two-stage MC simulation of both, aquifer uncertainty and population variability. The results indicate an exceedance of tolerable risk thresholds. Assuming a long-term occupancy in the area connected with usage of groundwater for household purposes, the cancer risk increase was calculated to range from ~+2% to ~+12%. Additionally the respective Hazard Quotients of all chloroethene compounds showed a spatially widespread exceedance of tolerable values. As a recommendation for the respective case study area, the exposure of humans to contaminated groundwater should be limited to a minimum. Regulatory measures are advisable to protect the affected population from harm. Additionally, indoor air measurements should be conducted to assess exposure via inhalation. This procedure appears reasonable due to the high volatility of chloroethenes, especially VC, and the high groundwater level of approximately 1 m below ground surface.

Regarding potential remediation measures the comprehension of redox conditions should be accomplished. As shown in this work, inorganic electron acceptors play a crucial role in degradation of chlorinated ethenes in the subsurface. In particular, it was demonstrated that degradation of cis-DCE only occurs if nitrate and sulphate are depleted (see chapter 4.2). Based on this observation in-situ remediation approaches like augmentation with carbon-containing substrates, e.g. EHC or molasses, should aim to be conducted in areas where nitrate reduction as well as sulphate reduction is already completed. The preceding evaluation of redox conditions assures that degradation-enhancing potential of the augmented substances is not wasted for reduction of inorganic electron acceptors.

8 Outlook and future prospects

Due to the physical properties of chlorinated ethenes, which lead to widespread contamination plumes (that are in most cases untreatable ex-situ at reasonable costs), much effort has been put into in-situ remediation techniques. One of these techniques found in literature, called bioaugmentation, uses addition of contaminant-degrading bacteria to contaminated soil. Different studies concerning bioaugmentation with the dechlorinating enrichment culture KB-1 (Major et al., 2002) or *Dehalococcoides* spp. (Schaefer et al., 2010) showed promising results in terms of contaminant decay improvement. Nevertheless, the mentioned literature also suggests highly reducing conditions for the enhancement of chloroethene degradation. As a consequence methods to lower sulphate concentrations, e.g. by reduction using methanol or acetate (Major et al., 2002), toluene, benzoate or lactate (Shen and Sewell, 2005) or other carbon substrates like EHC or glucose (chapter 4.2; Lee et al., 2004), seems to be a fundamental step prior to bioaugmentation. For the study area this work refers to, bioaugmentation along with addition of carbon substrates may be a feasible remediation action due to the fact that no bacteria capable of complete mineralisation of chloroethenes were detected by the PCR approach (chapter 4.3).

Another problem for modelling groundwater contaminations is the impossibility to directly measure degradation rates of contaminants in-situ. Due to the low decay rates, long-term time series of measurement data sets are necessary to assess degradation rates with conventional techniques. Even with the groundwater analyses available for the study area used in this work, the deduction of degradation rates was not possible due to fluctuation of concentration values. A recent method to evaluate degradation rates and assess self-healing potential of contaminated soils, e.g. for natural attenuation approaches, is called isotope fractionation. Isotope fractionation measurements are based on the presumption, that enzyme-mediated degradation reactions preferably catalyse reactions of molecules with incorporated ^{12}C isotopes rather than ^{13}C isotopes, which leads to an accumulation of ^{13}C-molecules in educt rather than product molecules ($^{13}C/^{12}C$ isotope fractionation ratio) (Ertl et al., 1996). This fractionation is detectable via gas chromatography coupled with isotope ratio mass spectrometry (GC-IRMS, Sherwood Lollar et al., 2001) and delivers valuable information on the prevailing degradation procedures (Vieth et al., 2003) or even in-situ reaction rates (Morrill et al., 2005).

Batch experiments (chapter 4.2) reveal that additional tests with different initial conditions are advisable. For instance, the postulated inhibition effect of inorganic electron acceptors (chapter 5.3.3; Widdowson, 2004) may be verified using this technique. In order to assess the inhibiting influence of both, manganese(IV) and iron(III) ions, an experimental setup with excess concentrations of these electron acceptors should be conducted in a similar way as performed in this work. The subsequent derivation of inhibition constants for these substances might help in predicting pollutant degradation in groundwater.

Additionally, a multi-pathway risk assessment approach which intertwines not only oral ingestion of contaminants, but also inhalation of indoor air and contact with contaminated soil should be tested for the study area in order to review alternative risk pathways. Different studies concerning this issue have already been performed (Chen and Ma, 2006; Ma, 2002), showing a major influence of ingestion of drinking water, but however, other pathways should not be neglected either. For the specific study site used in this work, the particular circumstances (i.e. high groundwater head and increasing concentrations of cis-DCE and VC) may lead to an increased risk via contaminant inhalation. Due to missing measurement data sets for validation purposes, the implementation of indoor air exposure to volatile organic contaminants was skipped in this work. However, application of model approaches available in literature (Whelan et al., 1992; McKone, 1993) using the simulation results of the present study may be a feasible method to evaluate risk.

For pending remediation actions planned for the mentioned site, a variety of options is available and described in literature (Khan et al., 2004; Stupp et al., 2007). Mainly in-situ techniques are applicable to the study site due to the large extent of the contamination plume. Kaufmann et al. (2005) estimate the remediation cost of DNAPLs in a sandy aquifer to reach up to 100,000 US-\$$\cdot$kg^{-1}. For similar pollutant scenarios, Stupp et al. (2006) calculated remediation costs for chlorinated ethenes using "pump and treat" measures to range from 2,125 – 4,000 €$\cdot$kg^{-1} if pollutant concentrations exceed 5000 µg$\cdot$l^{-1}. Lower pollutant concentrations come along with increased costs for remediation due to decreasing efficiency of the "pump and treat" method. Based on the precautious assumption, that the contaminant plume at the study site extends over 50,000 m^2 with an average thickness of 5 m and an estimated average concentration of 2 mg$\cdot$l^{-1}, about 500 kg of pollutants are distributed in the entire area within the fluid phase. Additionally, large amounts of contaminants are adsorbed onto soil particles and are not included in the calculation. Nevertheless, using the above given values for solute concentrations only, even the most cost-saving estimation of remediation costs results in a total price above 1,000,000 € for the given example.

With respect to cost-effectiveness, contaminant hot spots in the core area located in the upstream part of the model domain should be addressed for clean-up activities. However, as shown in this work, focus on the downstream region should also be considered due to the risk for the residents arising from the groundwater pollution. Less-impacted areas may be considered for a less cost-intensive monitored natural attenuation.

9 Nomenclature

9.1 Chemical compounds and trivial names

cis-DCE	1,2-cis-dichloroethene
CH_2O	(hypothetical) carbohydrate
EDTA	Ethylenediaminetetraacetic acid
EHC	mixture for in-situ groundwater remediation containing zero-valent iron and a carbon source
Fe(II)	iron ion (divalent)
Fe(III)	iron ion (trivalent)
Mn(II)	manganese ion (divalent)
Mn(IV)	manganese ion (tetravalent)
NO_3	nitrate ion
O_2	oxygen
PCE	tetrachloroethene
SO_4	sulphate ion
TCE	trichloroethene
VC	vinyl chloride

9.2 Latin symbols

a_{ox}, a_{red}	activity of relevant species of oxidant and reductant
C	concentration
D	damage/consequences of a certain event
$D_{L,k}$	longitudinal hydrodynamic dispersion of species k
$D_{d,k}$	coefficient of molecular diffusion of species k
$D_{T,k}$	transversal hydrodynamic dispersion of species k
E	reduction potential
E^0	standard reduction potential
e	gravitational unit vector
F	Faraday constant ($F = 9.648533 \cdot 10^4$ C·mol^{-1})
f_{OC}	fraction of organic carbon

ΔG^0	free Gibbs energy
h	hydraulic (piezometric) head
I	unit tensor
$K_r(s)$	relative hydraulic conductivity ($0 < Kr \leq 1$, $Kr = 1$ if saturated at $s = 1$)
K_f	tensor of hydraulic conductivity for the saturated medium (anisotropy)
$K_{D,k}$	partition coefficient of species k
K_{OW}	octanol-water partition coefficient
k_k	reaction rate constant for species k
$K_{S,k}$	half-saturation constant for species k
$K_{I,k}$	inhibition constants for species k
k_D	death rate
L	loading of sorbent
P	probability of occurrence
Q_k	zero-order non-reactive production term
q	Darcy flux vector
r_{max}	degradation rate
Ri	risk
R	universal gas constant ($R = 8.314472\ \mathrm{J \cdot K^{-1} \cdot mol^{-1}}$)
R_k	retardation factor
T	absolute temperature
z_e	number of transferred electrons

9.3 Greek symbols

$\alpha_L,\ \alpha_T$	longitudinal and transverse dispersivity of porous medium
ε	volume fraction / porosity
μ	mean value
μ_{max}	maximum specific growth rate
ν	reaction / degradation rate
ρ_s	soil dry density
σ	standard deviation
χ	buoyancy coefficient including fluid density effects

9.4 Subscripts

k	species indicator, $k = 1,..., n$
j	pathway indicator
meas	measured value
sim	simulated value
abs	absolute value
norm	normalised value
org	value in originating model
e-acc	electron-acceptor
MO	micro-organism

9.5 Abbreviations

AAS	Atomic Absorption Spectrometry
ADD	average daily dose
AT	averaging time
ATSDR	Agency of Toxic Substances and Disease Registry
BC	boundary condition
bp	base pair
BImSchG	Bundes-Immissionsschutzgesetz
BImSchV	Bundes-Immissionsschutzverordnung
BTEX	general term for benzene, toluene, ethylbenzene, xylene
BW	body weight
CHC	chlorinated hydrocarbons
CSF	cancer slope factor
CSTR	continuously stirred-tank reactor
DD	daily dose
DNA	Deoxyribonucleic acid
DNAPL	dense non-aqueous phase liquids
DOC	dissolved organic carbon
ED	exposure duration
EEA	European Environmental Agency
EF	exposure frequency

Err	quadratic deviation
GC/MS	Gas Chromatography coupled with Mass Spectrometry
GC-IRMS	Gas Chromatography coupled with Isotope Ratio Mass Spectrometry
GR	ingestion rate of tap water
HPLC	High Performance Liquid Chromatography
HQ	Hazard Quotient
MC	Monte-Carlo simulation
MNA	monitored natural attenuation
PAH	polyaromatic hydrocarbons
RfD	reference dose
rRNA	ribosomal Ribonucleic acid
SOM	soil organic matter
TOC	total organic carbon
U.S. EPA	United States Environmental Protection Agency
VOC	volatile organic compounds
VSS	volatile suspended solids

10 References

[1]   Alvarez-Cohen, L. and Speitel, G. E. (2001). Kinetics of aerobic cometabolism of chlorinated solvents. *Biodegradation, 12*(2), 105–126.

[2]   Amos, B., Christ, J., Abriola, L., Pennell, K. and Loeffler, F. (2007). Experimental evaluation and mathematical modeling of microbially enhanced tetrachloroethene (PCE) dissolution. *Environmental Science & Technology, 41*(3), 963–970.

[3]   Atkins, P. W., Trapp, C. A. and Höpfner, A. (2001). *Physikalische Chemie* (3rd edn). Weinheim: Wiley-VCH.

[4]   Azadpour-Keeley, A., Keeley, J. W., Russell, H. H. and Sewell, G. W. (2001). Monitored natural attenuation of contaminants in the subsurface: Processes. *Ground Water Monitoring & Remediation, 21*(2), 97–107.

[5]   Aziz, C. E., Newell, C. J. and Gonzales, J. (2000). *BIOCHLOR Chlorinated solvent plume database report*. San Antonio, Texas.

[6]   Ball, W. P. and Roberts, P. V. (1991). Long-term sorption of halogenated organic chemicals by aquifer material. 1. Equilibrium. *Environmental Science & Technology, 25*(7), 1223–1237.

[7]   Becker, J. (2006). A modeling study and implications of competition between Dehalococcoides ethenogenes and other tetrachloroethene-respiring bacteria. *Environmental Science & Technology, 40*(14), 4473–4480.

[8]   Benekos, I. D., Shoemaker, C. A. and Stedinger, J. R. (2007). Probabilistic risk and uncertainty analysis for bioremediation of four chlorinated ethenes in groundwater. *Stochastic Environmental Research and Risk Assessment, 21*(4), 375–390.

[9]   Biswas, N., Zytner, R. G., McCorquodale, J. A. and Bewtra, J. K. (1991). Prediction of the movement of perchloroethylene in soil columns. *Water, Air, & Soil Pollution, 60*(3), 361–380.

[10]   Bordeleau, G., Martel, R., Schäfer, D., Ampleman, G. and Thiboutot, S. (2008). Groundwater flow and contaminant transport modelling at an air weapons range. *Environmental Geology, 55*(2), 385–396.

[11]   Bradley, P. M. (2000). Microbial degradation of chloroethenes in groundwater systems. *Hydrogeology Journal, 8*(1), 104–111.

[12]   Bradley, P. M. and Chapelle, F. H. (2010). *Biodegradation of Chlorinated Ethenes* (In: In-Situ Remediation of Chlorinated Solvent Plumes, 1st edn). Berlin: Springer.

[13]   Briggs, G. (1981). Theoretical and experimental relationships between soil adsorption, octanol-water partition coefficients, water solubilities, bioconcentration factors, and the parachor. *Journal of agricultural and food chemistry, 29*(5), 1050–1059.

[14]   Chang, H. and Alvarez-Cohen, L. (1996). Biodegradation of individual and multiple chlorinated aliphatic hydrocarbons by methane-oxidizing cultures. *Applied and Environmental Microbiology, 62*(9), 3371.

[15]   Chang, J.-H., Qiang, Z. and Huang, C.-P. (2006). Remediation and stimulation of selected chlorinated organic solvents in unsaturated soil by a specific enhanced electrokinetics. *Colloids and Surfaces A: Physicochemical and Engineering Aspects, 287*(1-3), 86–93.

[16]   Chen, Y. and Ma, H. (2006). Model comparison for risk assessment: A case study of contaminated groundwater. *Chemosphere, 63*(5), 751–761.

[17]   Christensen, T., Bjerg, P., Banwart, S., Jakobsen, R., Heron, G. and Albrechtsen, H. (2000). Characterization of redox conditions in groundwater contaminant plumes. *Journal of Contaminant Hydrology, 45*(3-4), 165–241.

[18]   Clement, T. P., Sun, Y., Hooker, B. S. and Petersen, J. N. (1998). Modeling Multispecies Reactive Transport in Ground Water. *Ground Water Monitoring & Remediation, 18*(2), 79–92.

[19]   Clement, T. P., Truex, M. J. and Lee, P. (2002). A case study for demonstrating the application of U.S. EPA's monitored natural attenuation screening protocol at a hazardous waste site. *Journal of Contaminant Hydrology, 59*(1-2), 133–162.

[20]   Cupples, A., Spormann, A. and McCarty, P. (2004). Comparative evaluation of chloroethene dechlorination to ethene by Dehalococcoides-like microorganisms. *Environmental Science & Technology, 38*(18), 4768–4774.

[21]   Devlin, J. F., Katic, D. and Barker, J. F. (2004). In situ sequenced bioremediation of mixed contaminants in groundwater. *Journal of Contaminant Hydrology, 69*(3-4), 233–261.

[22]   Diersch, H.-J. (2006). *Feflow White Papers: Volume IV*. Berlin: DHI-WASY.

[23]   Diersch, H.-J. (2009). *Feflow Reference Manual*. Berlin: DHI-WASY.

[24]   DIN EN ISO 15680 (2003). Water quality - Gas-chromatographic determination of a number of monocyclic aromatic hydrocarbons, naphthalene and several chlorinated compounds using purge-and-trap and thermal desorption. Published in: 2004/04: DIN German Institute for Standardization, Berlin.

[25]   Dolfing, J., van Eeckert, M. and Mueller, J. (2006). Thermodynamics of low Eh reactions, *Proceedings of Battelle's 5th International Conference on Remediation of Chlorinated and Recalcitrant Compounds*. Monterey, California.

[26]   Doong, R., Wu, S. and Chen, T. (1996). Anaerobic biotransformation of polychlorinated methane and ethene under various redox conditions. *Chemosphere, 32*(2), 377–390.

[27]   Ertl, S., Seibel, F., Eichinger, L., Frimmel, F. H. and Kettrup, A. (1996). Determination of the 13C/12C Isotope Ratio of Organic Compounds for the Biological Degradation of Tetrachloroethene (PCE) and Trichloroethene (TCE). *Acta hydrochimica et hydrobiologica, 24*(1), 16–21.

[28]   European Environment Agency (2005). *The European Environment - State and outlook 2005*. Copenhagen.

[29]   Fan, C., Wang, G., Chen, Y. and Ko, C. (2009). Risk assessment of exposure to volatile organic compounds in groundwater in Taiwan. *Science of The Total Environment, 407*(7), 2165–2174.

[30]   Garabedian, S. P., LeBlanc, D. R., Gelhar, L. W. and Celia, M. A. (1991). Large-scale natural gradient tracer test in sand and gravel, Cape Cod, Massachusetts: 2. Analysis of spatial moments for a nonreactive tracer. *Water Resour. Res, 27*(5), 911–924.

[31]   Gelhar, L., Welty, C. and Rehfeldt, K. (1992). A critical review of data on field-scale dispersion in aquifers. *Water Resources Research, 28*(7), 1955–1974.

[32]   Gouy, M., Guindon, S. and Gascuel, O. (2010). SeaView version 4: a multiplatform graphical user interface for sequence alignment and phylogenetic tree building. *Molecular biology and evolution, 27*(2), 221.

[33]   Greis, T., Helmholz, K., Schöniger, H. and Haarstrick, A. (2011). Modelling of spatial contaminant probabilities of occurrence of chlorinated hydrocarbons in an urban aquifer. *Environmental Monitoring and Assessment*. DOI: 10.1007/s10661-011-2209-1. In press.

[34]   Haston, Z. and McCarty, P. (1999). Chlorinated ethene half-velocity coefficients (Ks) for reductive dehalogenation. *Environmental Science & Technology, 33*(2), 223–226.

[35]   Haws, N., Bouwer, E. and Ball, W. (2006). The influence of biogeochemical conditions and level of model complexity when simulating cometabolic biodegradation in sorbent-water systems. *Advances in Water Resources, 29*(4), 571–589.

[36]   Helmholz, K. (2011). *Numerical Stochastic Simulation for Remedial Activites and Risk Assessment for an Urban Groundwater System*. Doctoral Thesis. TU Braunschweig, Braunschweig. In progress.

[37]   Hendrickson, E. R., Payne, J. A., Young, R. M., Starr, M. G., Perry, M. P., Fahnestock, S., et al. (2002). Molecular Analysis of Dehalococcoides 16S Ribosomal DNA from Chloroethene-Contaminated Sites throughout North America and Europe. *Applied and Environmental Microbiology, 68*(2), 485–495.

[38]   Jang, W. and Aral, M. (2007). Modeling of co-existing anaerobic-aerobic biotransformations of chlorinated ethenes in the subsurface, *The Third International Conference on Environmental Science and Technology*. Houston, Texas.

[39]   Karickhoff, S. W., Brown, D. S. and Scott, T. A. (1979). Sorption of hydrophobic pollutants on natural sediments. *Water research, 13*(3), 241–248.

[40]   Kaufman, M. M., Rogers, D. T. and Murray, K. S. (2005). An empirical model for estimating remediation costs at contaminated sites. *Water, Air, & Soil Pollution, 167*(1), 365–386.

[41]   Kaufman, M., Rogers, D. and Murray, K. (2009). Using soil and contaminant properties to assess the potential for groundwater contamination to the lower Great Lakes, USA. *Environmental Geology, 56*(5), 1009–1021.

[42]   Khadam, I. and Kaluarachchi, J. J. (2003). Applicability of risk-based management and the need for risk-based economic decision analysis at hazardous waste contaminated sites. *Environment International, 29*(4), 503–519.

[43]   Khan, F. I., Husain, T. and Hejazi, R. (2004). An overview and analysis of site remediation technologies. *Journal of Environmental Management, 71*(2), 95–122.

[44]   Krajmalnik-Brown, R., Holscher, T., Thomson, I. N., Saunders, F. M., Ritalahti, K. M. and Loffler, F. E. (2004). Genetic Identification of a Putative Vinyl Chloride Reductase in Dehalococcoides sp. Strain BAV1. *Applied and Environmental Microbiology, 70*(10), 6347–6351.

[45]   LaBolle, E. and Fogg, G. (2001). Role of molecular diffusion in contaminant migration and recovery in an alluvial aquifer system. *Transport in Porous Media, 42*(1), 155–179.

[46]   Lee, I.-S., Bae, J.-H., Yang, Y. and McCarty, P. L. (2004). Simulated and experimental evaluation of factors affecting the rate and extent of reductive dehalogenation of chloroethenes with glucose. *Journal of Contaminant Hydrology, 74*(1-4), 313–331.

[47]   Lemke, L. and Bahrou, A. (2009). Partitioned multiobjective risk modeling of carcinogenic compounds in groundwater. *Stochastic Environmental Research and Risk Assessment, 23*(1), 27–39.

[48]   Ling, M. and Rifai, H. S. (2007). Modeling Natural Attenuation with Source Control at a Chlorinated Solvents Dry Cleaner Site. *Ground Water Monitoring & Remediation, 27*(1), 108–121.

[49]   Ma, H. (2002). Stochastic multimedia risk assessment for a site with contaminated groundwater. *Stochastic Environmental Research and Risk Assessment, 16*(6), 464–478.

[50]   Magnuson, J., Romine, M., Burris, D. and Kingsley, M. (2000). Trichloroethene reductive dehalogenase from Dehalococcoides ethenogenes: sequence of tceA and substrate range characterization. *Applied and Environmental Microbiology, 66*(12), 5141.

[51]   Maillard, J., Charnay, M.-P., Regeard, C., Rohrbach-Brandt, E., Rouzeau-Szynalski, K., Rossi, P., et al. (2011). Reductive dechlorination of tetrachloroethene by a stepwise catalysis of different organohalide respiring bacteria and reductive dehalogenases. *Biodegradation.*

[52]   Major, D. W., McMaster, M. L., Cox, E. E., Edwards, E. A., Dworatzek, S. M., Hendrickson, E. R., et al. (2002). Field demonstration of successful bioaugmentation to achieve

dechlorination of tetrachloroethene to ethene. *Environmental Science & Technology,* *36*(23), 5106–5116.

[53] Marin, C. M., Guvanasen, V. and Saleem, Z. A. (2003). The 3MRA Risk Assessment Framework — A Flexible Approach for Performing Multimedia, Multipathway, and Multireceptor Risk Assessments Under Uncertainty. *Human and Ecological Risk Assessment: An International Journal, 9*(7), 1655–1677.

[54] Massabo, M., Catania, F., Piazza, D. D. and Paladino, O. (2008). Groundwater risk Assessment Of PCE at a contaminated site. *Fresenius Environmental Bulletin, 17*(9).

[55] Maxwell, R. M. and Kastenberg, W. E. (1999). Stochastic environmental risk analysis: an integrated methodology for predicting cancer risk from contaminated groundwater. *Stochastic Environmental Research and Risk Assessment, 13*(1), 27–47.

[56] McKone, T. (1991). Human Exposure to Chemicals from Multiple Media and through Multiple Pathways: Research Overview and Comments. *Risk analysis, 11*(1), 5–10.

[57] McKone, T. E. (1993). *CalTOX, a multimedia total exposure model for hazardous-waste sites: Part 1, Executive summary.* Livermore, California.

[58] Michaelis, L. and Menten, M. L. (1913). Kinetics of invertase action. *Biochem. Z, 49*, 333–369.

[59] Miller, G. S., Milliken, C. E., Sowers, K. R. and May, H. D. (2005). Reductive Dechlorination of Tetrachloroethene to trans -Dichloroethene and cis -Dichloroethene by PCB-Dechlorinating Bacterium DF-1. *Environmental Science & Technology, 39*(8), 2631–2635.

[60] Mills, W., Cheng, J. J., Droppo Jr, J., Faillace, E., Gnanapragasam, E., Johns, R., et al. (1997). Multimedia benchmarking analysis for three risk assessment models: RESRAD, MMSOILS, and MEPAS. *Risk analysis, 17*(2), 187–201.

[61] Monod, J. (1949). The growth of bacterial cultures. *Annual Reviews in Microbiology, 3*(1), 371–394.

[62] Morrill, P. L., Lacrampe-Couloume, G., Slater, G. F., Sleep, B. E., Edwards, E. A., McMaster, M. L., et al. (2005). Quantifying chlorinated ethene degradation during reductive dechlorination at Kelly AFB using stable carbon isotopes. *Journal of contaminant hydrology, 76*(3-4), 279–293.

[63] Muller, J., Rosner, B., Abendroth, G. von, Meshulam-Simon, G., McCarty, P. and Spormann, A. (2004). Molecular identification of the catabolic vinyl chloride reductase from Dehalococcoides sp. strain VS and its environmental distribution. *Applied and Environmental Microbiology, 70*(8), 4880.

[64] Mulligan, C. N. and Yong, R. N. (2004). Natural attenuation of contaminated soils. *Environment International, 30*(4), 587–601.

[65] Noell, A. L. (2009). Estimation of Sequential Degradation Rate Coefficients for Chlorinated Ethenes. *Practice Periodical of Hazardous, Toxic, and Radioactive Waste Management, 13*(1), 35–44.

[66] Olaniran, A. O., Pillay, D. and Pillay, B. (2004). Chloroethenes contaminants in the environment: Still a cause for concern. *African Journal of Biotechnology, 3*(12).

[67] Piver, W., Jacobs, T. and Medina Jr, M. (1997). Evaluation of health risks for contaminated aquifers. *Environmental Health Perspectives, 105*(1), 127–143.

[68] Pliefke, T., Sperbeck, S. T., Urban, M., Peil, U. and Budelmann, H. (2007). A standardized methodology for managing disaster risk – An attempt to remove ambiguity. *Proceedings of the 5th International Probabilistic Workshop, Ghent. Belgium.*

[69] Robbins, G. A., Hayden, J. M. and Bristol, R. D. (1989). Vertical dispersion of ground water contaminants in the near-field of leaking underground gasoline storage tanks, *Petroleum Hydrocarbons and Chemicals in Groundwater Conference, November 15-17th.* Houston, Texas.

[70]   Ruffino, B. and Zanetti, M. (2009). Adsorption Study of Several Hydrophobic Organic Contaminants on an Aquifer Material. *American Journal of Environmental Sciences, 5*(4), 507–515.

[71]   Saiki, R., Scharf, S., Faloona, F., Mullis, K., Horn, G., Erlich, H., et al. (1985). Enzymatic amplification of beta-globin genomic sequences and restriction site analysis for diagnosis of sickle cell anemia. *Science, 230*(4732), 1350–1354.

[72]   Sangster Research Laboratories (May 4th, 2011). LogKOW: A databank of evaluated octanol-water partition coefficients (Log P). Canadian National Committee for CODATA. http://logkow.cisti.nrc.ca/logkow/index.jsp. Accessed July 27th, 2011.

[73]   Schaefer, C. E., Towne, R. M., Vainberg, S., McCray, J. E. and Steffan, R. J. (2010). Bioaugmentation for treatment of dense non-aqueous phase liquid in fractured sandstone blocks. *Environmental Science & Technology, 44*(13), 4958–4964.

[74]   Schaerlaekens, J., Mallants, D., Simûnek, J., van Genuchten, M. T. and Feyen, J. (1999). Numerical simulation of transport and sequential biodegradation of chlorinated aliphatic hydrocarbons using CHAIN_2D. *Hydrological Processes, 13*(17), 2847–2859.

[75]   Scheffer, F., Schachtschabel, P. and Blume, H.-P. (2008). *Lehrbuch der Bodenkunde* (15th edn). Heidelberg: Spektrum Akad. Verl.

[76]   Schwarzenbach, R. and Westall, J. (1981). Transport of nonpolar organic compounds from surface water to groundwater. Laboratory sorption studies. *Environmental Science & Technology, 15*(11), 1360–1367.

[77]   Schwarzenbach, R. P., Gschwend, P. M. and Imboden, D. M. (2003). *Environmental organic chemistry* (2nd edn). New York: Wiley.

[78]   Schwoerbel, J. (1987). *Einführung in die Limnologie* (6th edn, Vol. 31). Stuttgart: Fischer.

[79]   Shen, H. and Sewell, G. W. (2005). Reductive Biotransformation of Tetrachloroethene to Ethene during Anaerobic Degradation of Toluene: Experimental Evidence and Kinetics. *Environmental Science & Technology, 39*(23), 9286–9294.

[80]   Sherwood Lollar, B., Slater, G. F., Sleep, B., Witt, M., Klecka, G. M., Harkness, M., et al. (2001). Stable Carbon Isotope Evidence for Intrinsic Bioremediation of Tetrachloroethene and Trichloroethene at Area 6, Dover Air Force Base. *Environmental Science & Technology, 35*(2), 261–269.

[81]   Stegmann, N. (1969). *Entwicklung eines Darstellungsverfahrens für Baugrundkarten an Hand der Baugrundverhältnisse der Stadt Braunschweig.* Doctoral Thesis. Technische Universität Braunschweig, Braunschweig.

[82]   Stupp, H. and Paus, L. (1999). Migrationsverhalten organischer Grundwasser-Inhaltsstoffe und daraus resultierende Ansätze zur Beurteilung von Monitored Natural Attenuation (MNA). *TerraTech, Zeitschrift für Altlasten und Bodenschutz, 1999*(5), 32–37.

[83]   Stupp, H. D., Bakenhus, A., Stauffer, R. and Lorenz, D. (2005). Sanierungsoptimierung von CKW-Grundwasserschäden – Möglichkeiten zur Reduzierung der Sanierungskosten. *Altlasten Spektrum, 2005*(6), 313–322.

[84]   Stupp, H. D., Bakenhus, A., Stauffer, R. and Lorenz, D. (2006). Kosten zur Sanierung von Grundwasserverunreinigungen durch CKW und Ansätze zur Definition der Verhältnismäßigkeit von Sanierungsmaßnahmen. *Altlasten Spektrum, 1/2006*, 84–92.

[85]   Stupp, H., Bakenhus, A., Gass, M., Hüttmann, S. and Engelmann, F. (2007). Biologische Verfahren zur Sanierung von CKW-Grundwasserschäden – Systematik und Beschreibung der In-Situ-Techniken. *Altlasten Spektrum, 2007*(3), 101–110.

[86]   U.S. Department of Health and Human Services - Agency for Toxic Substances and Disease Registry (1996). *Toxicological Profile for 1,2-Dichloroethylene.* Atlanta, Georgia.

[87]   U.S. Department of Health and Human Services - Agency for Toxic Substances and Disease Registry (1997a). *Toxicological Profile for Tetrachloroethylene.* Atlanta, Georgia.

[88]   U.S. Department of Health and Human Services - Agency for Toxic Substances and Disease Registry (1997b). *Toxicological Profile for Trichloroethylene*. Atlanta, Georgia.

[89]   U.S. Department of Health and Human Services - Agency for Toxic Substances and Disease Registry (2006). *Toxicological Profile for Vinyl Chloride*. Atlanta, Georgia.

[90]   U.S. Environmental Protection Agency (1988). *Health Effects Assessment for Tetrachloroethylene: EPA/600/8-89-096*. Cincinnati, Ohio.

[91]   U.S. Environmental Protection Agency (2000). *Toxicological Review of Vinyl Chloride*. Washington, DC.

[92]   U.S. Environmental Protection Agency (2008). *Toxicological Review of Tetrachloroethylene (Perchloroethylene) - Draft*. Washington, DC.

[93]   U.S. Environmental Protection Agency (2009). *Toxicological Review of Trichloroethylene - Draft*. Washington, DC.

[94]   U.S. Environmental Protection Agency (2010). *Toxicological Review of cis-1,2-Dichloroethylene and trans-1,2-Dichloroethylene*. Washington, DC.

[95]   Umweltbundesamt (2010). Bundesweite Übersicht zur Altlastenstatistik. http://www.umweltbundesamt-daten-zur-umwelt.de/umweltdaten/public/document/downloadPrint.do?ident=19861. Accessed September 9th, 2011.

[96]   US EPA (1997). *MMSOILS: Multimedia Contaminant Fate, Transport, and Exposure Model: Documentation and User's Manual Version 4.0*. Washington, DC.

[97]   Valsaraj, K., Kommalapati, R., Robertson, E. and Constant, W. D. (1999). Partition constants and adsorption/desorption hysteresis for volatile organic compounds on soil from a Louisiana Superfund site. *Environmental Monitoring and Assessment, 58*(2), 227–243.

[98]   Vieth, A., Müller, J., Strauch, G., Kästner, M., Gehre, M., Meckenstock, R. U., et al. (2003). In-situ biodegradation of tetrachloroethene and trichloroethene in contaminated aquifers monitored by stable isotope fractionation. *Isotopes in environmental and health studies, 39*(2), 113–124.

[99]   Villemur, R., Saucier, M., Gauthier, A. and Beaudet, R. (2002). Occurrence of several genes encoding putative reductive dehalogenases in Desulfitobacterium hafniense/frappieri and Dehalococcoides ethenogenes. *Canadian journal of microbiology, 48*(8), 697–706.

[100] Watts, J., Fagervold, S., May, H. and Sowers, K. (2005). A PCR-based specific assay reveals a population of bacteria within the Chloroflexi associated with the reductive dehalogenation of polychlorinated biphenyls. *Microbiology, 151*(6), 2039.

[101] Whelan, G., Buck, J. W., Strenge, D. L., Droppo Jr, J. G., Hoopes, B. L. and Aiken, R. J. (1992). Overview of the multimedia environmental pollutant assessment system (MEPAS). *Hazardous Waste and Hazardous Materials, 9*(2), 191–208.

[102] Widdowson, M. A. (2004). Modeling natural attenuation of chlorinated ethenes under spatially varying redox conditions. *Biodegradation, 15*(6), 435–451.

[103] Wiedemeier, T. H., Swanson, M. A., Moutoux, D. E., Gordon, E. K., Wilson, J. T., Kampbell, D. H., et al. (1998). *Technical protocol for evaluating natural attenuation of chlorinated solvents in ground water*. Cincinnati, Ohio.

[104] Wilson, J. T., Kampbell, D. H., Ferrey, M. and Estuesta, P. (2001). *Evaluation of the Protocol for Natural Attenuation of Chlorinated Solvents: Case Study at the Twin Cities Army Ammunition Plant*. Washington, DC.

[105] Yan, J., Rash, B. A., Rainey, F. A. and Moe, W. M. (2009). Detection and Quantification of Dehalogenimonas and "Dehalococcoides" Populations via PCR-Based Protocols Targeting 16S rRNA Genes. *App Env Microbiol (Applied and Environmental Microbiology), 75*(23), 7560–7564.

[106] Yang, Q., Shang, H., Li, H., Xi, H. and Wang, J. (2008). Biodegradation of tetrachlorothylene using methanol as co-metabolic substrate. *Biomedical and Environmental Sciences, 21*(2), 98.

[107] Young, D. F. and Ball, W. P. (1998). Estimating Diffusion Coefficients in Low-Permeability Porous Media Using a Macropore Column. *Environmental Science & Technology, 32*(17), 2578–2584.

11 Appendix

11.1 Finite Element model and settings for the soil column experiments

Table 11.1: General column layout parameters for adsorption models

Dimension	Two-dimensional
Height	0.7 m
Diameter	0.072 m
Type	Saturated
Projection	Axisymmetric
Element-type	4-noded quadrilateral
Mesh elements	249
Mesh nodes	500
Problem class	Combined flow and mass transport
Time class	Transient flow – transient mass transport
Time stepping scheme	Forward Euler / Backward Euler
Upwinding	No Upwinding

Table 11.2: Temporal and control data for adsorption models

Time step control	Automatic (via predictor-corrector schemes)
Initial time step length	0.01 d
Maximum time step length	0.1 d
Initial time	0 d
Final time	0.7 d

Table 11.3: General flow data for adsorption models

Conductivity	$1e^{-5}$ m/s
Flow boudaries	
Head	0 m (at top of column)
Flux	-0.707 m/d (at bottom of column)

Table 11.4: Transport and sorption data for adsorption models

Porosity	0.3
Diffusion coefficient	0 m²/s
Transverse dispersivity	0 m
Longitudinal dispersivity	0.4 m
Sorption coefficients	
PCE	4.424
TCE	2.66
DCE	2.142
VC	1.43

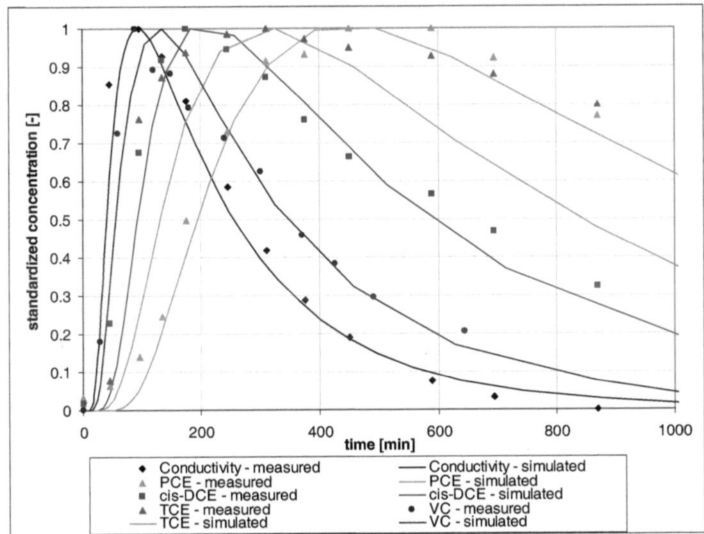

Figure 11.1: Adsorption experiment and simulation of chloroethenes in the soil column showing all 5 species (see chapter 4.1)

11.2 Procedures and techniques for characterisation of chloroethene-degrading bacteria (chapter 4.3)

DNA extraction protocol:

1. a) Centrifuge 50 ml of groundwater sample in a Falcon tube with 3000 rpm for 15 min. Transfer the pellet into a 2 ml tube

    b) Filtrate 400 ml of groundwater sample through a cellulose filter. Cut the filter in small pieces and transfer these into into a 2 ml tube

2. Add 400 mg glass beads and 1 ml of lysis buffer containing 100 mM Tris (pH 8), 100 mM EDTA, 100 mM NaCl, 1% (wt/vol) polyvinylpyrrolidone and 2% (wt/vol) sodiumdodecylsulphate

3. Homogenize the sample for 40 s at speed 6.0 in a cell disruptor

4. Centrifuge during 2 min at room temperature and 13,000 rpm

5. Extract with 1 volume phenol:chloroform, mix, centrifuge and take the supernatant

6. Extract with 1 volume chloroform, mix, centrifuge and take the supernatant

7. Collect the supernatant and add 1:10 volume of sodium acetate (3 M) and 1 volume isopropyl alcohol (-20℃)

8. Mix and incubate 20 min at -20℃. Centrifuge at 13,000 rpm during 30 min

9. Remove the supernatant and solve the pellet in 200 µl of TE buffer. In this step it is possible to incubate the sample at 65℃ in agitatio n (100 rpm) if needed for solving the pellet

10. (In meantime) equilibrate Sepharose 4B in a spin column with TE buffer. Transfer the sample to the Sepharose 4B and centrifuge 1 min at 13,000 rpm. Repeat with new Sepharose 4B if the sample is not colourless yet. DNA is passing through the matrix and the humic acids are adsorbed

11. Add 20 µl sodium acetate (1:10 volume) and 600 µl of ethanol (3x volume)

12. Incubate 20 min at -20℃ and centrifuge 30 min at 13,000 rpm

13. Remove the supernatant and solve the pellet in 50 µl pure water

14. Measure concentration of DNA taking 5 µl of sample and 70 µl of TE buffer: Spectrophotometer ratio 260 nm / 280 nm. Check the quality and size of the DNA by eletrophoresis on 1% agarose TAE gel. For further steps the concentration of DNA must be higher than 50-100 ng·µl^{-1}.

PCR protocol:

Table 11.5: Primer design for the PCR with groundwater samples

Primer	Sequence [5'->3']	GC content [%]	T$_{melt}$ [°C]	Fragment size
vcrA (forward)	TTGGGACGGTGGGTTACAC	58	58.8	300 bp
vcrA (reverse)	ATCTGCCAGGTTAAGAGCAC	50	57.3	
bvcA (forward)	CGGTTACAATCATCAGGGAG	50	57.3	295 bp
bvcA (reverse)	GGACATCATTTGCTGGGTAC	50	57.3	
tceA (forward)	GGACCCATGCTCTTGTATTG	50	57.3	260 bp
tceA (reverse)	CGCAGGTCTTACAGAACTC	53	56.7	
Dehalobacter (forward)	TACACCGACCTGGAACTTG	53	56.7	230 bp
Dehalobacter (reverse)	ACCGTTATAGGCCCAGAAAG	50	57.3	
16S-rRNA 728mod (forward)	GCGGTTTTCTAGGTTGTCAC	50	57.3	- 126 bp (with 792 reverse primer) - 435 bp (with 1172mod reverse primer)
16S-rRNA 774 (forward)	GGGAGTATCGACCCTCTC	61	58.2	- 328 bp (with 1172mod reverse primer)
16S-rRNA 792 (reverse)	ACAGAGAGGGTCGATACTC	53	56.7	See above
16S-rRNA 1172mod (reverse)	GTTTCGCGGGGCAGTCT	65	57.6	See above

Table 11.6: Composition of the PCR master mix

	Per sample (total 50 µl)	Per 10 samples (total 500 µl)
DNA template	2 µl	---
PCR buffer (5x)	10 µl	100 µl
dNTPs (10 nM)	5 µl	50 µl
Primers (forward + reverse)	1 + 1 µl	10 + 10 µl
Dest. Water	30.7 µl	307 µl
TAQ polymerase	0.3 µl	3 µl

PCR thermocycler settings:

- 94°C, 4 min
    - 94°C, 30 sec ⎤
    - 50°C, 40 sec ⎬ repeat 35 times
    - 72°C, 30 sec ⎦
- 72°C, 1 min
- 4°C, ∞

Transfer 4 µl of sample with 1 µl loading buffer into 2% (wt/vol) agarose gel and run the gel for 20 min at ~80 V. Gel staining is done by incubation in ethidium bromide solution (0.1% vol/vol) for 15 min.

11.3 Stoichiometric factors for degradation of chlorinated ethenes

Reaction equations for chlorinated ethene degradation:

1. $Cl_2C=CCl_2 + H_2 \rightarrow Cl_2C=CClH + H^+ + Cl^-$ PCE degradation
2. $Cl_2C=CClH + H_2 \rightarrow ClHC=CHCl + H^+ + Cl^-$ TCE degradation
3. $ClHC=CHCl + H_2 \rightarrow ClHC=CH_2 + H^+ + Cl^-$ DCE degradation
4. $ClHC=CH_2 + H_2 \rightarrow H_2C=CH_2 + H^+ + Cl^-$ VC degradation

Table 11.7: Chlorinated ethene degradation and stoichiometric matrix

	PCE	TCE	DCE	VC	Ethen	Cl^-	H_2	H^+
1. PCE degradation	-1	+0,79				+0,21	-0,012	+0,006
2. TCE degradation		-1	+0,74			+0,27	-0,015	+0,0076
3. DCE degradation			-1	+0,64		+0,37	-0,021	+0,0103
4. VC degradation				-1	+0,45	+0,57	-0,032	+0,016

Reaction equations for reduction of inorganic electron acceptors using a hypothetical carbohydrate:

1. $(CH_2O) + O_2 \rightarrow CO_2 + H_2O$ — Aerobic respiration
2. $(CH_2O) + 0,8\ H^+ + 0,8\ NO_3^- \rightarrow CO_2 + 1,4\ H_2O + 0,4\ N_2$ — Denitrification
3. $(CH_2O) + MnO_2 + 2\ H^+ \rightarrow CO_2 + H_2O + Mn^{2+} + H_2$ — Manganese reduction
4. $(CH_2O) + 0,67\ Fe_2O_3 + 1,33\ H^+ \rightarrow CO_2 + H_2O + 1,33\ Fe^{2+} + 0,67\ H_2$ — Iron reduction
5. $(CH_2O) + 0,83\ SO_4^{2+} + 1,67\ H^+ + 1,33\ H_2 \rightarrow CO_2 + 0,83\ H_2S + 2,33\ H_2O$ — Sulphate reduction

Table 11.8: Competing redox reactions involved in chlorinated ethene degradation

	CH	O_2	NO_3^-	Mn_4^+	Fe^{3+}	SO_4^{2+}	N_2	Mn^{2+}	Fe^{2+}	H_2S	CO_2	H_2O	H_2	H^+
1. Aerobic respiration	-1	-1,067									+1,47	+0,6		
2. Denitrification	-1		-1,653				+0,37				+1,47	+0,84		-0,027
3. Manganese reduction	-1			-2,9				+1,83			+1,47	+0,6	0,066	-0,066
4. Iron reduction	-1				-3,55				+2,49		+1,47	+0,6	0,044	-0,044
5. Sulphate reduction	-1					-2,67				+0,94	+1,47	+1,4	-0,088	-0,055

11.4 Brownfield site statistics for Germany, 2010

Table 11.9: Brownfield site statistics for Germany according to Umweltbundesamt (Date: 2010) (as discussed in chapter 2.3.1)

	date	suspicious areas	old waste deposits	abandoned industrial sites	brownfield sites	remediation completed	exposure assessment completed	currently remediated	monitored
Baden-Würtemberg	12/2009	14472	1968	12504	2124	2445	14312	635	413
Bayern	03/2010	16545	11450	5095	1084	1490	4590	1006	78
Berlin	07/2010	4978	1142	4468	911	187	---	68	75
Brandenburg	06/2010	19885	7140	12745	1454	3997	4327	127	214
Bremen	06/2010	3560	27	3533	432	596	898	43	170
Hamburg	07/2010	1876	272	1623	519	429	3024	135	139
Hessen	07/2010	1044	554	490	424	812	1624	181	40
Mecklenburg-Vorpommern	12/2009	5907	2678	3229	1049	1222	284	341	429
Niedersachsen	06/2010	99783	9399	90384	2948	1478	4095	360	325
Nordrhein-Westfalen	01/2010	75370	30493	44877	---	6158	17969	---	---
Rheinland-Pfalz	07/2010	12408	11947	461	294	127	6305	167	57
Saarland	05/2010	1977	1650	323	456	156	379	35	64
Sachsen	04/2010	20018	6799	13219	667	2836	6474	468	1393
Sachsen-Anhalt	05/2010	17210	5248	11962	173	1436	3271	74	27
Schleswig-Holstein	12/2009	13682	2092	11590	311	951	2585	69	42
Thüringen	03/2010	13583	4072	9511	814	739	4241	234	70
total		322298	96931	226014	13660	25059	74378	3943	3536

11.5 Measurement data from measurement campaigns at Schützenplatz area

Table 11.10a: Measurement data from groundwater wells including measurement uncertainty

	B12	SB1	B1	B5	B8	B10	B11	TUBS1
PCE [µg·l^{-1}]	4432 ± 2053	388 ± 284	5.7 ± 7.9	28 ± 31	59 ± 48	1990 ± 1062	18 ± 10	3 ± 6
TCE [µg·l^{-1}]	1586 ± 894	82 ± 48	0.3 ± 0.8	5 ± 7	15 ± 25	680 ± 666	1.1 ± 0.8	0 ± 0
cis-DCE [µg·l^{-1}]	6488 ± 2322	8245 ± 1478	3.8 ± 8.9	9 ± 17	26 ± 48	44 ± 44	10 ± 21	453 ± 356
VC [µg·l^{-1}]	6 ± 8	4 ± 6	0 ± 0	0 ± 0	0 ± 0.2	0 ± 0	0.1 ± 0.2	5 ± 6
pH value [-]	6.9 ± 0.1	7.0 ± 0.1	6.9 ± 0.1	6.8 ± 0.2	7.0 ± 0.1	6.82	7.3 ± 0.1	6.9 ± 0.1
Conductivity [mS·cm^{-1}]	1606 ± 262	1544 ± 136	1224 ± 132	1378 ± 115	1100 ± 101	1270	636 ± 254	1436 ± 142
redox potential [mV]	200 ± 54	13 ± 7	268 ± 49	230 ± 34	250 ± 38	-30	4.8 ± 28	-89 ± 21
diss. Oxygen [mg·l^{-1}]	0.8 ± 0.3	0.8 ± 0.4	1.2 ± 0.4	1.4 ± 0.3	1.1 ± 0.2	1.0	1.1 ± 0.7	0.9 ± 0.5
diss. organic carbon (DOC) [mg·l^{-1}]	5 ± 4	4 ± 4	3.4 ± 0.3	1.4 ± 0.9	6.5 ± 4.2	4.7	2.8 ± 3.9	7 ± 2
Nitrate [mg·l^{-1}]	11 ± 5	1.5 ± 2.4	15 ± 7.7	33 ± 10	16 ± 6	14	5.1 ± 3.1	0.4 ± 0.7
Nitrite [mg·l^{-1}]	0.5 ± 0.6	0.5 ± 0.7	0.5 ± 0.7	0.7 ± 1.2	0.3 ± 0.8	0	0.6 ± 1	0.5 ± 1.1
Chloride [mg·l^{-1}]	98 ± 26	99 ± 21	36 ± 13	61 ± 14	68 ± 13	177	99 ± 64	94 ± 20
Sulphate [mg·l^{-1}]	542 ± 113	491 ± 41	317 ± 42	492 ± 42	222 ± 31	243	10 ± 4	460 ± 31
Sodium [mg·l^{-1}]	97 ± 10	79 ± 17	33 ± 5	67 ± 14	56 ± 10		63 ± 24	59 ± 13
potassium [mg·l^{-1}]	22 ± 6	12 ± 4	24 ± 4	12 ± 3	12 ± 2		10 ± 1	12 ± 4
Magnesium [mg·l^{-1}]	35 ± 7	32 ± 5	30 ± 4	39 ± 5	14 ± 2		2 ± 1	43 ± 5
Calcium [mg·l^{-1}]	274 ± 47	239 ± 12	214 ± 25	243 ± 38	185 ± 33		64 ± 21	235 ± 39
Diss. Iron [mg·l^{-1}]	0.2 ± 0.2	0.2 ± 0.1	0.3	0.2 ± 0.2	0.4		0.1	13.8 ± 1.1
Diss. Manganese [mg·l^{-1}]	1.9 ± 0.4	1 ± 0.1	0	0.2 ± 0	0.6		1	2.9 ± 0.5

Table 11.10b: Measurement data from groundwater wells including measurement uncertainty

	B14	B15	B16	HA100	P1	RP9	RP15	P11
PCE [µg·l⁻¹]	90 ± 120	1686 ± 2339	4034 ± 2058	0.3 ± 0.6	57 ± 72	25 ± 14	25 ± 39	19 ± 34
TCE [µg·l⁻¹]	200 ± 630	438 ± 837	1012 ± 516	0 ± 0	12 ± 12	9 ± 7	6 ± 16	5 ± 13
cis-DCE [µg·l⁻¹]	9723 ± 5252	12884 ± 3064	759 ± 308	0.2 ± 0.4	192 ± 158	554 ± 478	8 ± 15	30 ± 27
VC [µg·l⁻¹]	26 ± 30	28 ± 29	6 ± 8	8 ± 15	4 ± 7	52 ± 31	0 ± 0	14 ± 15
pH value [-]	7.0 ± 0.1	7.0 ± 0.1	6.8 ± 0.2	7.0 ± 0.5	7.0 ± 0.1	6.8 ± 0.5	7.0 ± 0.1	7.0 ± 0.1
Conductivity [mS·cm⁻¹]	1540 ± 340	1592 ± 157	1551 ± 248	477 ± 137	1294 ± 90	1346 ± 235	1523 ± 177	1401 ± 146
redox potential [mV]	-32 ± 57	-63 ± 34	206 ± 96	20 ± 43	-74 ± 39	-82 ± 42	-84 ± 22	-96 ± 22
diss. Oxygen [mg·l⁻¹]	0.6 ± 1	0.8 ± 0.4	0.7 ± 0.4	1 ± 1.4	0.9 ± 0.3	1.1 ± 0.4	0.9 ± 0.4	0.8 ± 0.3
diss. organic carbon (DOC) [mg·l⁻¹]	4.3 ± 2.6	4.5 ± 3.0	4.9 ± 4.1	9.8 ± 1.7	3.5 ± 2.1	4.7 ± 3.2	5.2 ± 5.1	6.7 ± 7.9
Nitrate [mg·l⁻¹]	0.1 ± 0.7	0.3 ± 0.6	20 ± 9	2.6 ± 1.1	0.8 ± 1.6	0.5 ± 1.1	0.4 ± 0.7	0.9 ± 0.9
Nitrite [mg·l⁻¹]	0.1 ± 2.7	0.2 ± 0.6	0.3 ± 0.6	0.8 ± 1.1	0.5 ± 1	0.3 ± 0.5	0.3 ± 0.9	0.4 ± 0.8
Chloride [mg·l⁻¹]	101 ± 26	89 ± 18	99 ± 19	33 ± 8	63 ± 4	87 ± 49	74 ± 23	65 ± 7
Sulphate [mg·l⁻¹]	513 ± 142	654 ± 73	544 ± 108	74 ± 40	457 ± 56	520 ± 216	366 ± 103	433 ± 97
Sodium [mg·l⁻¹]	86 ± 22	78 ± 9	121 ± 40	26 ± 7	57 ± 8	57 ± 18	78 ± 13	63 ± 10
potassium [mg·l⁻¹]	10 ± 2	5 ± 2	17 ± 3	6 ± 1	9 ± 2	6 ± 2	20 ± 4	7 ± 4
Magnesium [mg·l⁻¹]	35 ± 10	31 ± 3	28 ± 8	7 ± 3	39 ± 5	27 ± 8	68 ± 26	48 ± 7
Calcium [mg·l⁻¹]	229 ± 61	322 ± 35	274 ± 59	43 ± 38	232 ± 37	222 ± 55	211 ± 67	232 ± 41
Diss. Iron [mg·l⁻¹]	1.3 ± 0.8	4.9 ± 4.1	0.1 ± 0.2	20	13 ± 1		5 ± 3	18 ± 2
Diss. Manganese [mg·l⁻¹]	1.4 ± 0.5	2.3 ± 0.1	1.5 ± 0.2	4.4	2.3 ± 0.2		4.1 ± 0	2.7 ± 0.1

Table 11.10c: Measurement data from groundwater wells including measurement uncertainty

	GWM1	GWM2	GWM3	GWM4	GWM5	TUBS2	TUBS3	RP7
PCE [$\mu g \cdot l^{-1}$]	980 ± 384	8075 ± 464	7933 ± 3153	2300 ± 500	282 ± 362	1 ± 3	1 ± 1	4517 ± 2304
TCE [$\mu g \cdot l^{-1}$]	125 ± 67	2525 ± 591	2300 ± 346	863 ± 133	45 ± 48	0 ± 0	0 ± 0	1727 ± 928
cis-DCE [$\mu g \cdot l^{-1}$]	540 ± 304	1238 ± 1015	1143 ± 860	993 ± 737	15 ± 18	1 ± 3	3700 ± 1273	340 ± 244
VC [$\mu g \cdot l^{-1}$]	0.4 ± 0.6	3 ± 5	3 ± 4	31 ± 53	1 ± 1	0 ± 1	115 ± 21	3 ± 5
pH value [-]						7.0 ± 0.1	7.0 ± 0.3	6.8 ± 0.1
Conductivity [$mS \cdot cm^{-1}$]					1150	1419 ± 143	1117 ± 396	1433 ± 182
redox potential [mV]					235	-105 ± 44	-60 ± 28	197 ± 75
diss. Oxygen [$mg \cdot l^{-1}$]		1	0.9		0.9	0.8 ± 1.3	0.3 ± 0	0.7 ± 0.4
diss. organic carbon (DOC) [$mg \cdot l^{-1}$]	6.5 ± 3	38 ± 55	43 ± 62	42 ± 46	4.1	8.1 ± 3.9	2.4	4.9 ± 3.9
Nitrate [$mg \cdot l^{-1}$]		9	11		7	2 ± 3	1.8 ± 1.8	24 ± 14
Nitrite [$mg \cdot l^{-1}$]		0	0		0	0.3 ± 0.5	0.3 ± 0.5	0.2 ± 0.6
Chloride [$mg \cdot l^{-1}$]	48 ± 14	71 ± 36	57 ± 1.4	115	158	100 ± 10	85 ± 78	105 ± 30
Sulphate [$mg \cdot l^{-1}$]	197 ± 55	595 ± 97	529 ± 174	490 ± 85	359	418 ± 57	276 ± 161	439 ± 53
Sodium [$mg \cdot l^{-1}$]		81	69		58	68 ± 9	61 ± 28	108 ± 28
potassium [$mg \cdot l^{-1}$]		13	16		26	14 ± 5	4 ± 0	18 ± 3
Magnesium [$mg \cdot l^{-1}$]		32	30		28	48 ± 13	18 ± 11	26 ± 3
Calcium [$mg \cdot l^{-1}$]		322	277		252	199 ± 28	99 ± 24	214 ± 62
Diss. Iron [$mg \cdot l^{-1}$]		1.2	0.4			16 ± 6	8.6	0 ± 0
Diss. Manganese [$mg \cdot l^{-1}$]		2.6	4.5			4.2 ± 1	1.6	0.5 ± 0.1

11.6 Groundwater and soil analytics

Chlorinated ethene analysis:

Chlorinated ethene measurements were performed by the company Biolab Umweltanalysen GmbH according to DIN EN ISO 15680 by means of gas-chromatographic analysis using purge&trap and thermal desorption (DIN EN ISO 15680, 2003).

Cation and anion HPLC analysis:

Cation and anion concentrations in the groundwater and batch-test samples were analysed by High Performance Liquid Chromatography (HPLC) using a VWR-Hitachi LaChrom Elite system. The HPLC system was equipped with an electric conductivity detector (Merck Hitachi L-3720). An isocratic method was used for determination of both, anion and cation concentrations. Standard solutions for the analyses were prepared from pure substances (purchased from Carl Roth, Sigma-Aldrich, Fluka).

The following settings were used for analysis of anions:

- Column: Hamilton PRP-X110S
- Eluent: 1.7 mM $NaHCO_3$ (= 0.1428 $g \cdot l^{-1}$)

  1.8 mM Na_2CO_3 (= 0.1907 $g \cdot l^{-1}$)

  0.1 mM Na-thiocyanate (= 12.5 $\mu l \cdot l^{-1}$)

  in aqueous solution
- Flow: 2 ml/min
- Oven temperature: 30°C

The following settings were used for analysis of cations:

- Column: Alltech Cation
- Eluent: 3 $mM \cdot l^{-1}$ Methanesulfonic acid in aqueous solution
- Flow: 1 $ml \cdot min^{-1}$
- Oven temperature: 30°C

Iron and manganese AAS analysis:

Concentrations of iron and manganese were determined by atomic absorption spectrometry (AAS) using a Philips PU9200 AAS. A mixture of acetylene/air (70%/30%) with a flow of 0.9 $l \cdot min^{-1}$ was used for combustion of samples. An injection volume of 100 μl was used; measurements were performed in quadruple determination. All samples were at first filtered through a 0.2 µm filter and afterwards acidified using small volumes of concentrated nitric acid. Standard solutions purchased from Carl Roth were used for calibration purposes.

Detection of the different species was done with compound-specific lamps. Iron was detected at 248.3 nm, 0.2 nm bandpass, while manganese detection was performed at 279.5 nm, 0.5 nm bandpass.

Organic carbon analysis:

Quantification of the dissolved carbon (DC) and the dissolved inorganic carbon (DIC) content of leachate samples was conducted applying catalytic oxidation at 850 °C for DC and acidic catalysis at 160 °C for DIC. The dissolved organic carbon (DO C) content is calculated by differentiation of DC and DIC content. A carbon analyser (Dimatoc-2000) was used to measure the DC and DIC content in the liquid samples by determining carbon dioxide quantitatively by infrared absorption of carbon dioxide. All samples were filtered through a 0.2 µm filter before analysis. Standards were prepared with potassium hydrogen phthalate.

Measurement of dissolved organic carbon (DOC) in soil samples with a carbon analyser (DIMA-1000). TC content was measured by determining carbon dioxide quantitatively by thermal pulping at 1200 °C and infrared absorption of carbon dioxid e. Prior to the analysis, samples were oven-dried at 60 °C until constant weight.

11.7 Basic model parameters

General settings:

Table 11.11: General settings of the FE model "Schützenplatz"

Dimension	Three-dimensional
Extend	~760 x ~760 m
Layers	10
Type	saturated
Aquifer	unconfined
Element-type	6-noded triangular prisms
Mesh elements	273940
Mesh nodes	153043
Problem class	Combined flow and mass transport
Time class	Steady flow – transient mass transport
Time stepping scheme	Forward Euler / Backward Euler
Upwinding	No Upwinding

Table 11.12: Temporal and control data of the FE model "Schützenplatz"

Time step control	Automatic (via predictor-corrector schemes)
Initial time step length	0.01 d
Error tolerance	$1 \cdot 10^{-5}$
Initial time	0 d
Final time	25550 d

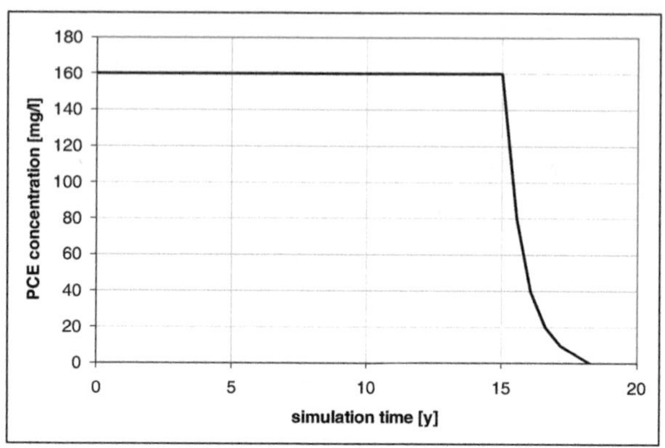

Figure 11.2: Dirichlet boundary condition used for infiltration of PCE

Settings for the reference model, MC simulation and optimisation approach:

Table 11.13: Hydraulic parameters of the reference model (chapter 5.1) and range of stochastic variation for the MC simulation (chapter 5.2)

Layer	Soil type	Hydraulic conductivity [e-4 ms^{-1}]	Porosity [-]	Longitudinal Dispersion [m]	Transversal Dispersion [m]
1	rubble	5 ± 1.25	0.25 ± 0.063	13.2 ± 3.3	1.98 ± 0.5
2	silt	1.16 ± 0.29	0.05 ± 0.013	9 ± 2.25	1.35 ± 0.34
3	fine sand	4.63 ± 1.16	0.1 ± 0.025	12 ± 3	1.8 ± 0.45
4	medium sand	23.15 ± 5.79	0.12 ± 0.03	11.4 ± 2.85	1.71 ± 0.43
5	fine sand	4.63 ± 1.16	0.1 ± 0.025	12 ± 3	1.8 ± 0.45
6	coarse sand	30 ± 7.5	0.17 ± 0.043	15 ± 3.75	2.25 ± 0.56
7	medium sand	23.15 ± 5.79	0.12 ± 0.03	11.4 ± 2.85	1.71 ± 0.43
8	fine sand	4.63 ± 1.16	0.1 ± 0.025	12 ± 3	1.8 ± 0.45
9	coarse sand	30 ± 7.5	0.17 ± 0.043	15 ± 3.75	2.25 ± 0.56
10	silt	1.16 ± 0.29	0.05 ± 0.013	9 ± 2.25	1.35 ± 0.34

Table 11.14: Species related parameters of the reference model (chapter 5.1) and stochastic variation for the MC simulation (chapter 5.2)

Compound	K_D-value [e-3]	Reaction rate [s^{-1}]	Diffusion coefficient [m²s^{-1}]
PCE	4.44 ± 1.11	1e-8 ± 3.75e-9	5e-8 ± 2.5e-9
TCE	4.44 ± 1.11	1e-8 ± 3.75e-9	5e-8 ± 2.5e-9
cis-DCE	3.89 ± 0.97	1e-9 ± 1.88e-10	5e-8 ± 2.5e-9
VC	3.33 ± 0.83	1e-8 ± 3.75e-9	5e-8 ± 2.5e-9

11.8 Parameter of the redox-dependant "Schützenplatz" model (chapter 5.3.2)

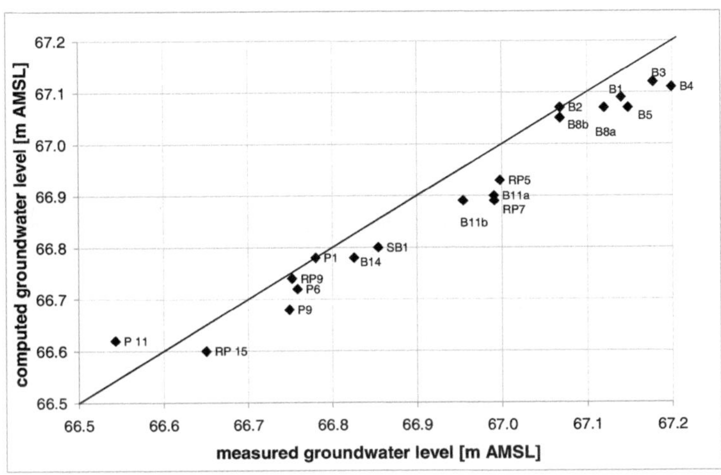

Figure 11.3: Scatter plot of measured and simulated water table of the preliminary reactive transport model

Table 11.15: Hydraulic parameters used in the redox-dependent degradation model (chapter 5.3.2)

Layer	Soil type	Hydraulic conductivity [e-4 m·s^{-1}]	Porosity [-]	Longitudinal Dispersion [m]	Transversal Dispersion [m]
1	rubble	0.5	0.175	15	5
2	silt	0.01	0.175	15	5
3	fine sand	3	0.175	15	5
4	medium sand	9	0.175	15	5
5	fine sand	3	0.175	15	5
6	coarse sand	18	0.175	15	5
7	medium sand	9	0.175	15	5
8	fine sand	3	0.175	15	5
9	coarse sand	18	0.175	15	5
10	silt	0.01	0.175	15	5

Table 11.16: Species related parameters used in the redox-dependent degradation model (chapter 5.3.2)

Compound	K_D-value [e-3]	Reaction rate [s^{-1}]	Diffusion coefficient [m$^2 \cdot$s^{-1}]
PCE	2.22	8e-8	5e-8
TCE	2.22	1.4e-7	5e-8
cis-DCE	0.056	1.2e-7	5e-8
VC	0.056	1.4e-7	5e-8

Table 11.17: Inhibition constants used in the redox-dependent degradation model (chapter 5.3.2)

Compound	Inhibition constant
Nitrate	0.1
Sulphate	10
PCE	5
TCE	1
cis-DCE	1

11.9 Parameters of the transect model (chapter 5.3.3)

Table 11.18: General model layout for transect model simulations

Dimension	Two-dimensional
Height	650 m
Diameter	50 m
Type	Saturated
Projection	Horizontal
Element-type	4-noded quadrilateral
Mesh elements	16250
Mesh nodes	16626
Problem class	Combined flow and mass transport
Time class	Transient flow – transient mass transport
Time stepping scheme	Forward Euler / Backward Euler
Upwinding	No Upwinding

Table 11.19: Temporal and control data for transect model simulations

Time step control	Automatic (via predictor-corrector schemes)
Initial time step length	0.001 d
Error tolerance	$1 \cdot 10^{-4}$
Initial time	0 d
Final time	18250 d

Table 11.20: General flow data for adsorption models

Conductivity	$5 \cdot 10^{-4}$ m·s^{-1}
Flow boudaries	
Head (top)	0 m
Head (bottom)	0.78 m

Table 11.21: Transport and sorption data for adsorption models

Porosity	0.278
Diffusion coefficient	$2.87e\text{-}8\ m^2 \cdot s^{-1}$
Transverse dispersivity	11 m
Longitudinal dispersivity	3.5 m
Sorption coefficients	[e-3]
PCE	3.89
TCE	3.33
DCE	0.67
VC	0.0056

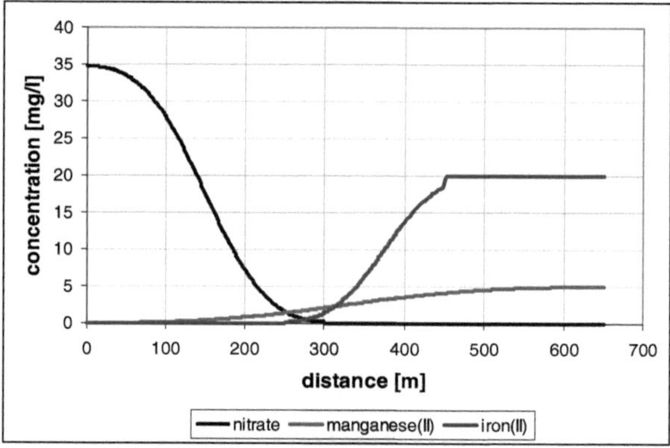

Figure 11.4: Concentration of electron acceptors over the transect length implemented in the transect model; fixed values (not shown) for oxygen (0.1 mg·l⁻¹) and sulphate (400 mg·l⁻¹)

Table 11.22: Inhibition constants of PCE degradation by different electron acceptors

	degradation by aerobic MOs	degradation by nitrate MOs	degradation by manganese MOs	degradation by iron MOs	degradation by sulphate MOs
Nitrate	10	---	---	---	---
Manganese	20	20	---	---	---
iron	30	30	30	---	---
sulphate	230	210	190	140	---

Table 11.23: Inhibition constants of TCE degradation by different electron acceptors

	degradation by aerobic MOs	degradation by nitrate MOs	degradation by manganese MOs	degradation by iron MOs	degradation by sulphate MOs
Nitrate	10	---	---	---	---
Manganese	20	15	---	---	---
iron	40	25	50	---	---
sulphate	270	130	150	150	---

Table 11.24: Inhibition constants of cis-DCE degradation by different electron acceptors

	degradation by aerobic MOs	degradation by nitrate MOs	degradation by manganese MOs	degradation by iron MOs	degradation by sulphate MOs
Nitrate	2	---	---	---	---
Manganese	3	5	---	---	---
iron	10	20	15	---	---
sulphate	50	50	50	50	---

Table 11.25: Inhibition constants of VC degradation by different electron acceptors

	degradation by aerobic MOs	degradation by nitrate MOs	degradation by manganese MOs	degradation by iron MOs	degradation by sulphate MOs
Nitrate	20	---	---	---	---
Manganese	20	20	---	---	---
iron	40	40	40	---	---
sulphate	400	400	300	150	---

Table 11.26: Growth and death rate for different microorganisms; growth inhibition constants for different electron acceptors

Microorganism species	Growth rate [10^{-4} s^{-1}]	Death rate [10^{-4} s^{-1}]	Growth inhibition constant [mg]			
			oxygen	nitrate	manganese	iron
Aerobic MO	0.5	0.002	---	---	---	---
Nitrate MO	0.4	0.002	1	---	---	---
Manganese MO	0.1	0.002	1	20	---	---
Iron MO	0.05	0.002	1	20	20	---
Sulphate MO	0.02	0.002	1	20	20	30

Band 1 **Sunder, Matthias**: Oxidation grundwasserrelevanter Spurenverunreinigungen mit Ozon und Wasserstoffperoxid im Rohrreaktor. 1996. FIT-Verlag Paderborn, ISBN 3-932252-00-4

Band 2 **Pack, Hubertus**: Schwermetalle in Abwasserströmen: Biosorption und Auswirkung auf eine schadstoffabbauende Bakterienkultur. 1996. FIT-Verlag Paderborn, ISBN 3-932252-01-2

Band 3 **Brüggenthies, Antje**: Biologische Reinigung EDTA-haltiger Abwässer. 1996. FIT-Verlag Paderborn, ISBN 3-932252-02-0

Band 4 **Liebelt, Uwe**: Anaerobe Teilstrombehandlung von Restflotten der Reaktivfärberei. 1997. FIT-Verlag Paderborn, ISBN 3-932252-03-9

Band 5 **Mann, Volker G.**: Optimierung und Scale up eines Suspensionsreaktorverfahrens zur biologischen Reinigung feinkörniger, kontaminierter Böden. 1997. FIT-Verlag Paderborn, ISBN 3-932252-04-7

Band 6 **Boll, Marco**: Einsatz von Fuzzy-Control zur Regelung verfahrenstechnischer Prozesse. 1997. FIT-Verlag Paderborn, ISBN 3-932252-06-3

Band 7 **Büscher, Klaus**: Bestimmung von mechanischen Beanspruchungen in Zweiphasenreaktoren. 1997. FIT-Verlag Paderborn, ISBN 3-932252-07-1

Band 8 **Burghardt, Rudolf**: Alkalische Hydrolyse – Charakterisierung und Anwendung einer Aufschlußmethode für industrielle Belebtschlämme. 1998. FIT-Verlag Paderborn, ISBN 3-932252-13-6

Band 9 **Hemmi, Martin**: Biologisch-chemische Behandlung von Färbereiabwässern in einem Sequencing Batch Process. 1999. FIT-Verlag Paderborn, ISBN 3-932252-14-4

Band 10 **Dziallas, Holger**: Lokale Phasengehalte in zwei- und dreiphasig betriebenen Blasensäulenreaktoren. 2000. FIT-Verlag Paderborn, ISBN 3-932252-15-2

Band 11 **Scheminski, Anke**: Teiloxidation von Faulschlämmen mit Ozon. 2001. FIT-Verlag · Paderborn, ISBN 3-932252-16-0

Band 12 **Mahnke, Eike Ulf**: Fluiddynamisch induzierte Partikelbeanspruchung in pneumatisch gerührten Mehrphasenreaktoren. 2002. FIT-Verlag Paderborn, ISBN 3-932252-17-9

Band 13 **Michele, Volker**: CFD modeling and measurement of liquid flow structure and phase holdup in two- and three-phase bubble columns. 2002. FIT-Verlag Paderborn, ISBN 3-932252-18-7

Band 14 **Wäsche, Stefan**: Einfluss der Wachstumsbedingungen auf Stoffübergang und Struktur von Biofilmsystemen. 2003. FIT-Verlag Paderborn, ISBN 3-932252-19-5

Band 15 **Krull Rainer**: Produktionsintegrierte Behandlung industrieller Abwässer zur Schließung von Stoffkreisläuren. 2003. FIT-Verlag Paderborn, ISBN 3-932252-20-9

Band 16 **Otto, Peter**: Entwicklung eines chemisch-biologischen Verfahrens zur Reinigung EDTA enthaltender Abwässer. 2003. FIT-Verlag Paderborn, ISBN 3-932252-21-7

Band 17 **Horn, Harald**: Modellierung von Stoffumsatz und Stofftransport in Biofilmsystemen. 2003. FIT-Verlag Paderborn, ISBN 3-932252-22-5

Band 18 **Mora Naranjo, Nelson**: Analyse und Modellierung anaerober Abbauprozesse in Deponien. 2004. FIT-Verlag Paderborn, ISBN 3-932252-23-3

Band 19 **Döpkens, Eckart**: Abwasserbehandlung und Prozesswasserrecycling in der Textilindustrie. 2004. FIT-Verlag Paderborn, ISBN 3-932252-24-1

Band 20 **Haarstrick, Andreas**: Modellierung millieugesteuerter biologischer Abbauprozesse in heterogenen problembelasteten Systemen. 2005. FIT-Verlag Paderborn, ISBN 3-932252-27-6

Band 21 **Baaß, Anne-Christina**: Mikrobieller Abbau der Polyaminopolycarbonsäuren Propylendiamintetraacetat (PDTA) und Diethylentriaminpentaacetat (DTPA). 2004. FIT-Verlag Paderborn, ISBN 3-932252-26-8

Band 22 **Staudt, Christian**: Entwicklung der Struktur von Biofilmen. 2006. FIT-Verlag Paderborn, ISBN 3-932252-28-4

Band 23 **Pilz, Roman Daniel**: Partikelbeanspruchung in mehrphasig betriebenen Airlift-Reaktoren. 2006. FIT-Verlag Paderborn, ISBN 3-932252-29-2

Band 24 **Schallenberg, Jörg**: Modellierung von zwei- und dreiphasigen Strömungen in Blasensäulenreaktoren. 2006. FIT-Verlag Paderborn, ISBN 3-932252-30-6

Band 25 **Enß, Jan Hendrik**: Einfluss der Viskosität auf Blasensäulenströmungen. 2006. FIT-Verlag Paderborn, ISBN 3-932252-31-4

Band 26 **Kelly, Sven**: Fluiddynamischer Einfluss auf die Morphogenese von Biopellets filamentöser Pilze. 2006. FIT-Verlag Paderborn, ISBN 3-932252-32-2

Band 27 **Grimm, Luis Hermann**: Sporenaggregationsmodell für die submerse Kultivierung koagulativer Myzelbildner. 2006. FIT-Verlag Paderborn, ISBN 3-932252-33-0

Band 28 **León Ohl, Andrés**: Wechselwirkungen von Stofftransport und Wachstum in Biofilsystemen. 2007. FIT-Verlag Paderborn, ISBN 3-932252-34-9

Band 29 **Emmler, Markus**: Freisetzung von Glucoamylase in Kultivierungen mit *Aspergillus niger*. 2007. FIT-Verlag Paderborn, ISBN 3-932252-35-7

Band 30 **Leonhäuser, Johannes**: Biotechnologische Verfahren zur Reinigung von quecksilberhaltigem Abwasser. 2007. FIT-Verlag Paderborn, ISBN 3-932252-36-5

Band 31 **Jungebloud, Anke**: Untersuchung der Genexpression in *Aspergillus niger* mittels Echtzeit-PCR. 1996. FIT-Verlag Paderborn, ISBN 978-3-932252-37-2

Band 32 **Hille, Andrea**: Stofftransport und Stoffumsatz in filamentösen Pilzpellets. 2008. FIT-Verlag Paderborn, ISBN 978-3-932252-38-9

Band 33 **Fürch, Tobias**: Metabolic characterization of recombinant protein production in *Bacillus megaterium*. 2008. FIT-Verlag Paderborn, ISBN 978-3-932252-39-6

Band 34 **Grote, Andreas Georg**: Datenbanksysteme und bioinformatische Werkzeuge zur Optimierung biotechnologischer Prozesse mit Pilzen. 2008. FIT-Verlag Paderborn, ISBN 978-3-932252-40-120

Band 35 **Möhle, Roland Bernhard**: An Analytic-Synthetic Approach Combining Mathematical Modeling and Experiments – Towards an Understanding of Biofilm Systems. 2008. FIT-Verlag Paderborn, ISBN 978-3-932252-41-9

Band 36 **Reichel, Thomas**: Modelle für die Beschreibung das Emissionsverhaltens von Siedlungsabfällen. 2008. FIT-Verlag Paderborn, ISBN 978-3-932252-42-6

Band 37 **Schultheiss, Ellen**: Charakterisierung des Exopolysaccharids PS-EDIV von *Sphingomonas pituitosa*. 2008. FIT-Verlag Paderborn, ISBN 978-3-932252-43-3

Band 38 **Dreger, Michael Andreas**: Produktion und Aufarbeitung des Exopolysaccharids PS-EDIV aus *Sphingomonas pituitosa*. 1996. FIT-Verlag Paderborn, ISBN 978-3-932252-44-0

Band 39   **Wiebels, Cornelia**: A Novel Bubble Size Measuring Technique for High Bubble Density Flows. 2009. FIT-Verlag Paderborn, ISBN 978-3-932252-45-7

Band 40   **Bohle, Kathrin**: Morphologie- und produktionsrelevante Gen- und Proteinexpression in submersen Kultivierungen von *Aspergillus niger*. 2009. FIT-Verlag Paderborn, ISBN 978-3-932252-46-2

Band 41   **Fallet, Claas**: Reaktionstechnische Untersuchungen der mikrobiellen Stressantwort und ihrer biotechnologischen Anwendungen. 2009. FIT-Verlag Paderborn, ISBN 978-3-932252-47-1

Band 42   **Vetter, Andreas**: Sequential Co-simulation as Method to Couple CFD and Biological Growth in a Yeast. 2009. FIT-Verlag Paderborn, ISBN 978-3-932252-48-8

Band 43   **Jung, Thomas**:   Einsatz chemischer Oxidationsverfahren zur Behandlung industrieller Abwässer. 2010. FIT-Verlag·Paderborn, ISBN 978-3-932252-49-5

Band 44   **Appel, Christina**: Numerische Simulation von Bioreaktorströmungen. 2010. FIT-Verlag Paderborn, ISBN: 978-3-932252-53-2

Band 45   **Herrmann, Tim**: Transport von Proteinen in Partikeln der Hydrophoben Interaktions Chromatographie. 2010. FIT-Verlag Paderborn, ISBN 978-3-932252-51-8

Band 46   **Becker, Judith**: Systems Metabolic Engineering of *Corynebacterium glutamicum* towards improved Lysine Production. 2010. Cuvillier-Verlag Göttingen, ISBN 978-3-86955-426-6

Band 47   **Melzer, Guido**: Metabolic Network Analysis of the Cell Factory *Aspergillus niger*. 2010. Cuvillier-Verlag Göttingen, ISBN 978-3-86955-456-3

Band 48   **Bolten J., Christoph**: Bio-based Production of L-Methionine in *Corynebacterium glutamicum*. 2010. Cuvillier-Verlag Göttingen, ISBN 978-3-86955-486-0

Band 49   **Lüders, Svenja**: Prozess- und Proteomanalyse gestresster Mikroorganismen. 2010. Cuvillier-Verlag Göttingen, ISBN 978-3-86955-435-8

Band 50   **Wittmann, Christoph**: Entwicklung und Einsatz neuer Tools zur metabolischen Netzwerkanalyse des industriellen Aminosäure-Produzenten *Corynebacterium glutamicum*. 2010. Cuvillier-Verlag Göttingen, ISBN 978-3-86955-445-7

Band 51   **Edlich, Astrid**: Entwicklung eines Mikroreaktors als Screening-Instrument für biologische Prozesse. 2010. Cuvillier-Verlag Göttingen, ISBN 978-3-86955-470-9

Band 52   **Hage, Kerstin**: Bioprozessoptimierung und Metabolomanalyse zur Proteinproduktion in *Bacillus licheniformis*. 2010. Cuvillier-Verlag Göttingen, ISBN 978-3-86955-578-2

Band 53   **Kiep, Katina Andrea**: Einfluss von Kultivierungsparametern auf die Morphologie und Produktbildung von *Aspergillus niger*. 2010. Cuvillier-Verlag Göttingen, ISBN 978-3-86955-632-1

Band 54   **Fischer, Nicole**: Experimental investigations on the influence of physico-chemical parameters on anaerobic degradation in MBT residual waste. 2011. Cuvillier-Verlag Göttingen, ISBN 978-3-86955-679-6

Band 55   **Schädel, Friederike**: Stressantwort von Mikroorganismen. 2011. Cuvillier-Verlag Göttingen, ISBN 978-3-86955-746-5

Band 56   **Wichter, Johannes**: Untersuchung der L-Cystein-Biosynthese in *Escherichia coli* mit Techniken der Metabolom- und ^{13}C-Stoffflussanalyse. 2011. Cuvillier-Verlag Göttingen, ISBN 978-3-86955-750-2

Band 57   **Knappik, Irena Isabell**: Charakterisierung der biologischen und chemischen Reaktionsprozesse in Siedlungsabfällen. 2011. Cuvillier-Verlag Göttingen, ISBN 978-3-86955-760-1

Band 58   **Driouch, Habib**: Systems biotechnology of recombinant protein production in *Aspergillus niger*. 2011. Cuvillier-Verlag Göttingen, ISBN 978-3-86955-808-0

Band 59   **Gehder, Matthias**: Development and Validation of Indicators for the Production and Quality of Seed Cultures. 2011. Cuvillier-Verlag Göttingen, ISBN 978-3-86955-847-9

Band 60   **Sommer, Becky**: Methodenentwicklung zur Charakterisierung sporenbildender Pilz-Seedingkulturen. 2011. Cuvillier-Verlag Göttingen, ISBN 978-3-86955-851-6

Band 61   **Dohnt, Katrin**: Charakterisierung von *Pseudomonas aeruginosa*-Biofilmen in einem *in vitro*-Harnwegskathetersystem. 2011. Cuvillier-Verlag Göttingen, ISBN 978-3-86955-852-3